Alachua County ARES
Operation Steinhatchee Storm
October 7, 2017

Simulated Emergency Test Planning and Execution: Steinhatchee Storm

Including
After Action Report / Improvement Plan
Approved, Nov 8, 2017

Volunteer Susan Halbert pounding out messages to transfer over the microwave link

Copyright © 2017 Gordon L. Gibby MD

All rights reserved except: Amateur radio emergency volunteer groups may copy as needed for improving their groups.

ISBN-13: 978-1978441507
ISBN-10: 1978441509

CONTENTS

	Acknowledgments	v
1	Background & Overview	1
2	Our Exercise Plan	5
3	The Nuts & Bolts of Planning	21
4	After Action Report: Exercise Overview	29
5	After Action Report: Exercise Design Summary	32
6	After Action Report: Analysis of Capabilities	38
7	After Action Report: Conclusion	45

Appendix A: Issues Noted / Improvement Plan	47
Appendix B: Selected ICS Forms	49
ICS 201 Situation	
Appendix C: ICS 205A Addendum	52
Appendix D: Lessons Learned	60
Appendix E: Participant Feedback Form (Suggested for Future Exercises)	61
Appendix F: Exercise Events Summary Table	64
Appendix G: Acronyms	65
Appendix H: RMS_RELAY SETTING	66

The Alachua County ARES Steinhatchee Storm Volunteers

Comm Unit #2 at Hungry Howies

ACKNOWLEDGMENTS

The Alachua County ARES group would like to acknowledge all the people who helped us make this growth possible, particularly all our own volunteers who worked so hard for so many months, and also the Alachua County EOC which has supported our efforts at backup emergency communications by purchasing multiple radios and large storage batteries. We would specifically like to acknowledge the gracious permission provided by The Crapps Family in allowing us to use the Jonesboro Lookout Tower.

Vann Chesney AC4QS at Comm Unit #1, Good Times, overlooking Steinhatchee River

The microwave system at COMM UNIT #2, Hungry Howies

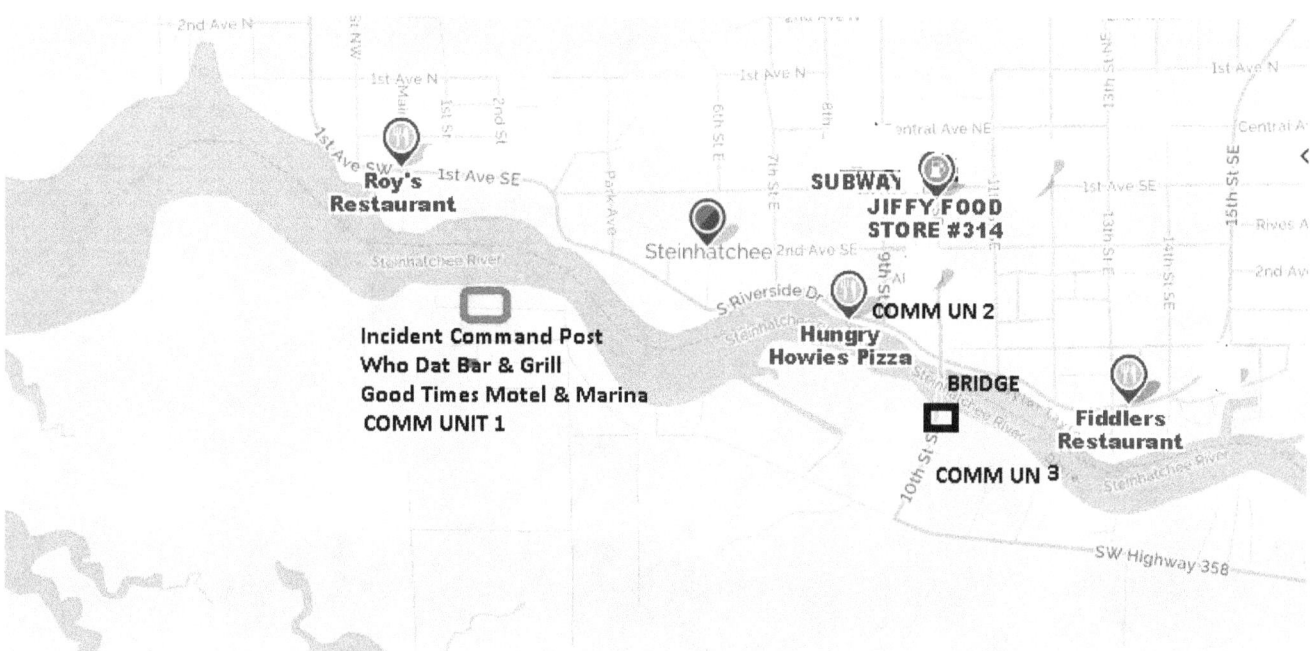

Locations of COMM UNIT 1, 2, and 3

Comm Unit 3 just before teardown (Normally that door was closed!)

(blank page)

SECTION 1: BACKGROUND & OVERVIEW

The purpose of this text is to record how our small Amateur Radio Emergency Service volunteer group carried out our 2017 Steinhatchee Storm Full Scale Exercise, in hopes that our successes and failures will benefit other groups who wish to carry out Full Scale Exercises.

Steinhatchee Storm 2017 was our group's second Full Scale Exercise, and was planned at a time that our group had undergone a substantial metamorphosis from a loosely trained organization of primarily VHF walkie-talkie volunteers, into a moderately-well-trained group with skills and assets in VHF voice, VHF packet, FLDIGI, and WINLINK. This drastic change took well over a year of enthusiastic leadership. During that year members of our group, sometimes individually and often as a group, installed 18 antennas as we developed a series of eight VHF linbpq nodes and other systems to create a real digital amateur radio emergency safety net. Training sessions were held at least monthly and often twice or three times monthly, as members were constantly given friendly encouragement to up their skills, assets, and strategies.

The group began with a core of perhaps 5-7 individuals who met monthly and planned the traditional "Fair Weather Fun" footrace- and other very useful and community-service oriented events, and also staffed a radio room at the local Emergency Operations Center and a couple of hurricane shelters during the occasional summer/fall hurricanes

All this new training resulted in many changes in the group, as some members found the direction and pace a bit much for them, and others were exhilarated. New Technician license classes, spurred by a demonstration at a CERT training, brought new members, and other hams begin to hear of exciting things happening at the ARES group. During this time period, the first local General/Extra Class license class was held, and five people ended up passing the Extra Class (highest license available) --- some of whom had started within this group, taking one of the original Technician courses! Internal growth was building.

Mentorship for the leaders of this group came from all over the world. Steve Waterman, Executive Director of the WINLINK organization, was a constant source of new challenges to grow. Other members of the WINLNK Development Team, including Mike Burton, provided additional technical expertise when we got lost. Incredibly patient John Wiseman G8BPQ, provided incredible Raspberry-Pi software and every-day-of-the-week help whenever we had questions. Dave Welker W2SRP of neighboring Marion County mentored us out of his multi-year experience running a thriving digitally-oriented Hospital Emergency Communications group.

It was Dave who challenged the author to take ICS-120 course online, on how to create Full Scale

Exercises, and that really "lit the fuse" and propelled our group into running the most rigorous communications exercises possible. After all, amateur radio communications is really most needed when all other means of communication are in trouble, so why not plan for the worst case scenario?

Our group carried out a 4-hour Hurricane-based Full Scale Exercise in May with surprise changes every hour, and a heavy emphasis on digital communications of all kinds --- a disaster for the first hour and then skills began to make sense and real growth happened in the team.

The Full Scale Exercise discussed in this text was planned to be a much more "laid back" exercise, but with new twists for our group --- an actual deployment miles and miles away from home; the use for the first time of a portable WINLINK RMS_RELAY / RMS_TRIMODE high-frequency server system to communicate with our VHF Raspberry Pi-based linbpq (John Wiseman) systems --- and for the first time, Ubiquity/AREDN based microwave systems.

We planned this exercise beginning early in the Summer of 2017, after our May Hurricane Test. Planning was accomplished well before the Puerto Rico/ Virgin Islands hurricane catastrophe, but since there are only so many different kinds of amateur radio emergency communications, we ended up planning trials of basically the same – or even more advanced – kinds of amateur radio communications techniques that are, at the time of this writing, still being used daily by the ARRL-American Red Cross "Force of 50" volunteers, and by an additional group of 10 SHARES volunteers who have just reached Puerto Rico. That makes our Full Scale Exercise obviously relevant to the cutting edge of Ham Radio Emergency Communications! Our group is literally practicing exactly the techniques that are literally forming the core of the largest amateur radio emergency response in history.

WINLINK has played a strong role in shaping our growth – the author started out in digital ham radio literally using a vacuum tube Heathkit SB-102 to learn PSK31 and WINLINK EXPRESS, and moved on to become a Sysop of an automated RMS-RELAY forwarding station that is daily moving Puerto Rico recovery traffic. Others in our group are Sysops of VHF WINLINK gateways, gaining expertise. Our group mentored a nearby amateur WD4SEN who was already well into this, and now his automated station is moving very large quantities of Puerto Rico traffic as well.

Following the HSEEP format (https://www.fema.gov/media-library-data/20130726-1914-25045-8890/hseep_apr13_.pdf ; template for reports: https://emergency.cdc.gov/training/ERHMScourse/pdf/127961885-Hseep-AAR-IP-Template-2007.pdf) for creating Full Scale Exercises, you must first outline the **Core Capabilities** you wish your group to possess; and the **Objectives** of your Exercise toward furthering your competency at those Core Capabilities. *Set those down in writing* --- they will be used over and over.

Next comes the creation of the **Exercise Plan**. Our final version of our written Exercise Plan is given in the next Section. The components of your Exercise Plan will force you to address issues of Safety, Security, Communications, and Schedules and Maps. MAPS turn out to be incredibly important when your group is deploying to unfamiliar surroundings --- so spend quite a bit of time making out maps that give people a firm grasp of where they are going to set up and what the circumstances will be. I employed multiple Internet resources, including satellite photos of the terrain, and even ground-level views at times to plan where setup could occur.

In the process of creating our Exercise Plan we had to figure out how to conduct a very long VHF digital link – 60 miles. An entire day of scouting out forestry lookout towers went into that effort, with photographs of each possibility. Thus I discovered a tower near to Steinhatchee Florida which preliminary testing revealed could by itself directly reach one of our existing digital nodes on 145.070. That set the stage for developing a VHF portion of our exercise (and ultimately was an unexpected failure during the actual Full Scale Exercise).

We wanted to gain further experience with ICS forms and systems during this Exercise and I had just recently completed ICS-300. (Several others in our group already had this training.) So we literally created every ICS form that would apply to our Exercise, which is a time-consuming and tedious work. In the end, we didn't use these to any huge extent --- but our volunteers, some of whom took on internal roles such as Safety Officer, Incident Commander, Unit Leader, Logistics Chief etc, learned a great deal about the practical workings of the Incident Command System that will make us better team-players when interfacing with extensively trained government officials.

Several permissions had to be gained to allow this Exercise to proceed. Those are mentioned in Section 4.

For our group, taking on new technologies every few months, training on new technologies HAS to happen early, well before the Full Scale Exercise. Exercises are not the time to learn the technology; they are times to learn how to USE the technology and better strategies HOW to use it. Thus for two months prior to the Full Scale Exercise, we demonstrated the new microwave technology, with working systems (as best as we were able to arrange as we constantly improved those systems).

We always hold a "**tabletop drill**" before a Full Scale Exercise. Meeting at a large house, we set up all the different teams in different rooms, and practice the technologies and the actual (or similar) events of the planned Full Scale Exercise. In our limited experience, these are usually a disaster! Things that should have worked simply don't and we spend a lot of time learning during these Tabletops. But with the next team just a few feet away, we can talk to each other and solve the problems a lot more easily than in the field. We literally write down EVERY failure that we experience during these Tabletops, and then we meticulously solve those failures BEFORE the subsequent Full Scale Exercise. There will be enough problems during the Full Scale Exercise to deal with --- better to solve all the KNOWN problems beforehand!

ICS Documents: In an Appendix, certain portions of our ICS documents are presented. A complete set of our ICS Documents for Steinhatchee Storm may be reviewed at our website: http://www.qsl.net/nf4rc/ A problem that we discovered is that some of the "stock" ICS forms are protected in such a way that you can't pull out just the important parts and leave behind the Instructions; and you can't grab the information part and insert it into word processing documents very easily if at all. That made our planning more difficult, but probably servers higher purposes in protecting the integrity of basic ICS forms.

In my experience, an e-mailing list, and a website are very important to communicating detailed information. Our volunteers are accustomed to a barrage of emails from me and somehow learn how to pick the ones of highest importance to them; my goal is to make certain that no volunteer feels "left out" or doesn't feel "out of the loop." We also have weekly Thursday-night VHF voice FM (repeater-based) short nets, which also provide a way to communicate information, and some (not all) of our group get together every other Saturday morning for breakfast. It is easy for any ham radio-related group to get a free web page at *www.qsl.net*, and their KOMPOZER free web site creation tool works acceptably well for creating an information-heavy web site. I have used the Coffee Cup free FTP tool, as well as the Open Office word processing tools. Thus, along with google mail, every one of our internet tools and resources has been obtained for the total cost of $0.00

Repetition. You will undoubtedly observe (and quickly tire of) that in ICS / HSEEP formats, the same material gets repeated several times. The same explanations and maps and objectives will be seen both in the Exercise Plan, ICS documents, and the After Action Report. That seems to be because each document needs to stand on its own.

After Action Report / Improvement Plan (AAR/IP)   NF4RC   2017 Steinhatchee Storm

SECTION 2 OUR EXERCISE PLAN

Version 1.02 Sept 5, 2017

SUB-SECTIONS

Title	Page #
	(deleted)
Exercise Overview	
Foreword	
Exercise scope, objectives, and core capabilities	
Participant roles and responsibilities	
Rules of conduct	
Safety Issues (including real emergency phrase)	
Logistics	
Security	
Communications	
Schedules	
Maps and Directions	

FOREWORD

Welcome!

Thank you so very much for participating in our training exercise! Giving of your time and preparation to benefit your community and neighbor are honorable and worthy actions --- and we appreciate your commitment and dedication.

This manual is designed to be read to give you an introduction to our upcoming Exercise – read it first, and then read the ICS-style documents that are being produced for our Exercise as well.

This manual is produced in accordance with the document **Homeland Security Exercise and Evaluation Program (HSEEP), April 2013**, which can be accessed here:

   https://www.fema.gov/media-library-data/20130726-1914-25045-8890/hseep_apr13_.pdf

In order to best understand these plans and exercises, it is recommended that the participant take free online courses at the ICS-100, 200, 700 and 800 introductory level. Addition coursework such as the ICS-300 and ICS-400 are also helpful. There are ICS courses on exercise planning that will help the participant learn more about the development process as well.

Thank you again for your participation! You're a real asset to your community!

Gordon L. Gibby MD
Newberry, Florida
August, 2018

EXERCISE SCOPE, OBJECTIVES, AND CORE CAPABILITIES

The Goal of the Alachua County Communications Plan
(broader than just one hurricane test goal)

TO FURNISH EMERGENCY COMMUNICATIONS WHEN REGULAR COMMUNICATIONS FAIL OR ARE INADEQUATE IN THE EVENT OF NATURAL OR MAN-MADE DISASTERS

The Goals of The 2017 EXERCISE STEINHATCHEE STORM Full Scale Exercise

PURPOSE
This exercise is designed to provide feedback on our proficiency and capabilities to achieve the likely communications tasks required in a severe weather emergency of sufficient magnitude to overwhelm or temporarily disable normal communications and normal grid power. It is also a learning opportunity for peripherally involved amateur radio operators, and local government and NGO personnel to become more aware of the abilities as well as the limitations of the local amateur radio ARES group.

CORE CAPABILITIES

Alachua County ARES Core Capabilities (as listed May, 2017)	Status	Involvement in this exercise
1. Antenna Placement	Significant growth of our group; 18 antennas emplaced and drills held to practice emergency antenna placement.	This exercise will test HF and VHF deployment and for the first time, microwave antenna deployment.
2. Emergency Simplex Repeater	Utilized successfully in May 2017 Test; deficiencies recognized, and not yet corrected.	A peripheral involvement in this test.
3. WINLINK Communications	Significant skills demonstrated in May 2017 Hurricane Exercise, but uneven skills across group.	Significant involvement in this S.E.T. exercise.
4. Backup Power	Several new "go-boxes" have been	Significant involvement in this

	built since the May 2017 Hurricane Test and more ARES members have backup power capabilities, but still not 100%	exercise
5. Mobile Deployment	Tested during May 2017 Hurricane Test, but not to the extent we would be testing it during the S.E.T.	Very significant involvement --- group will be deployed to a small city 60 miles away.
6. MT63 Skills	Rudimentary level of skills at present; not successfully tested in the May 2017 Hurricane Test.	Not involved in this exercise.
7. Packet Chat	Although the linbpq functions have not proven fast enough for wide usage, Vann Chesney has spearheaded getting "unconnected packet" chat skills developed in our group	Not involved in this exercise
8. LINBPQ Chat Functions	Acceptable only for up to 3 participants at this time.	Not involved in this exercise.
9. ICS Forms	Moderately developed skills that were somewhat utilized in the May 2017 Hurricane Test but need considerable improvement.	Intensive immersion in the ICS system and paperwork is planned

	OBJECTIVES
1	Assess the capabilities of our group to work within the Incident Command System framework on a deployed mission outside of Alachua County.
2	Assess the capability of our group to provide a 60+ mile digital VHF communications link, and provide practice to our members in making multiple connections to reach a distant station.
3	Provide practice for, and assess our capabilities at sending/receiving WINLINK email messages and attachments by VHF and/or HF and/or Microwave technologies.

SCOPE --- in multiple aspects

DIMENSION	LIMITATIONS
Kinds of Exercise Participants	Primarily ARES-associated licensed amateur radio operators with prior training but flexible enough to allow untrained amateur radio operators whom we are mentoring to participate; EOC personnel from the Emergency Manager's office where possible.

Geographic Area	Physical locations spread from Gainesville all the way to water's edge in Steinhatchee.
Number of Participants	Not any real limitation on the number of amateur radio operators, other than adding additional layers as needed for span of control issues.
Responder Functions	Communication of messages by VHF/HF/Microwave is the bottom line outcome, with process measurement of intermediate functions to achieve the end-goal of communications.
Hazard Type	Severe weather event, flooding, grounded cruise ship, with loss of normal cell phone / telephone / Internet / 700-800 MHz police/fire systems, and widespread loss of conventional power.

PARTICIPANT ROLES AND RESPONSIBILITIES

Participants will integrate into a mythical Incident Command System already in operation to deal with the weather disaster at Steinhatchee, Florida, and then assume new duties as that ICS expands to handle an evolving situation.

Our volunteers will staff multiple positions:

- Incident Commander
- Safety Officer
- Logistics Section Chief
  - Communications Unit #1
  - Communications Unit #2
  - Communications Unit #3
  - Communications Unit #4
- Planning Section Chief
- Operations Section Chief

Scenario

This is the approximate scenario we will be carrying out. The exact details (hours, tactics etc) may change slightly as the planning continues and as the incident plays out and the on-scene Incident Commander and Section Chiefs order, but this will serve to give the basic scenario.

Hurricane Thoughtless has been moving through the Gulf of Mexico for 2 days and finally settled in on a course that took it right to Steinhatchee, Florida, a sleepy fishing, tourist-scalloping town of about 1200 residents that has been flooded many times by hurricanes.

As the projected path of the Hurricane became more confident, about half the population of Steinhatchee voluntarily evacuated inland, remembering the recent flooding brought about by Hurricane Hermine that shut down large parts of the town for weeks and months. The Sheriff of Taylor County recognized the developing situation and activated the ICS system, becoming the Incident Commander, with careful communications to the Sheriff of Dixie County (beginning on the southern edge of the Steinhatchee River) as well as the town leadership of Steinhatchee, Keaton Beach (11 miles to the north) and Horseshoe Beach (to the south).

The Incident Commander took preparatory steps to have citizens prepare for the oncoming hurricane in the evening hours of October 6th, as winds were rising. Water was stored as much as possible, debris secured, boats secured as much as posssible, and everyone battened down the hatches for the coming storm, which was strengthening into a Category II hurricane.

Unbeknownst to the Incident Commander, cruise ship PoorNavigator was being driven by the storm and moving in the direction of their area, driven with the storm surge that allowed its 19-foot draft to enter the relatively shallow waters of the northern Gulf of Mexico. A bit behind the curve, the Captain of the cruise ship did not recognize the full implications of his loss of control of his track; similar to the tragic course of the barge El Faro that intentionally held course right through a hurricane a few years back, with the sinking and loss of all the crew.

As the storm strengthened, the Incident Commander grew increasingly concerned about the possible damages, which might be greater than what had been experienced by Steinhatchee in 2016. When the storm made landfall at 0400, not only did all electrical power go out (quite as expected) but also within 30 minutes, normal telecommunications, including cell phone, landline, and Internet also quit. Communications were still possible for another half hour with the remotely operated Coast Guard Marine Radio tower in Horseshoe Beach, but by 0500 internet control of that radio failed and the Incident Commander realized that his communications extended only as far as his marine FM transceiver could reach --- not a happy feeling. He decided to send a courier inland to try and ask for assistance. At 0600 a courier was able to leave the area, making a circuitous route to avoid the peak intensity of the inland hurricane, and found working telephone service at Old Town, Florida, and contacted State of Florida Emergency Management in Tallahassee. A request for some form of "longer-distance communications support" was communicated since normally Steinhatchee residents simply endure hurricanes to the best of their ability, but the Incident Commander did want the ability to request additional resources quickly should something unexpected happen.

The State EOC was watching the storm now turn toward Tallahassee and quite concerned about communications losses in the capital city that would need large amounts of cellphone capacity, but he was aware of a volunteer communications ARES group in Alachua County and reached out to the Alachua County EOC to see if they would be able to reach the area before larger satellite trucks could move in from caches throughout the Southeast. The Alachua County emergency manager contacted Jeff Capehart of the Alachua County ARES group (a part of ESF#2) and Jeff's response was eager and enthusiastic. As winds were really not dangerous in Gainesville, he knew a good group of communicators would be having their traditional Saturday breakfast at 0730. He thought he could muster a significant set of resources by 0830 or earlier to head westward, as the storm was weakening and sharply turned toward Tallahassee.

Realizing that the Alachua County ARES group has specific digital capabilities involving email, the Alachua County EOC agreed to monitor a prescribed email address for communications, in addition to remaining available on the 146.82 repeater and 146.52 simplex. The State EOC also indicated that they would also monitor a prescribed email address, but that they currently did not have packet amateur radio access.

Thus, by 0830 a caravan of volunteers headed westward from Gainesville toward the Check-In station at Casey's Cove just outside Steinhatchee, with multiple VHF, HF and even microwave gear. Permission to use the antenna on top of the Jonesboro lookout tower had been graciously provided by its private owner, who had purchased the tower years ago from the Florida Forest Service. The team thought that not only were HF connections quite easily made with portable stations, but VHF digital packet communications could be achieved by placing a portable "node" station at the Jonesboro lookout tower, and another portable "node" station near the highest point in town --- the bridge over the Steinhatchee River.

Upon reaching the Casey's Cove check-in station (a gas station/deli/convenience store with a large parking lot) the Incident Commander at the Good Times Marina bar and grill ("WATS DAT") was notified by FM marine radio.

Participants will fill out the ICS – 211 Incident Check-In List

Teams were given copies of the 0900 ICS 201 Incident Briefing, and of the 0900 Incident Action Plan, including multiple forms. Instructions (ICS form 204) were quickly given to emplace units at

- Incident Command Post, Good Times Marina
- Logistics station at Hungry Howies
- high ground at the FL-358 bridge over the Steinhatchee River
- Jonesboro lookout tower at the intersection of FL-358 and US-19

and the Communications Teams quickly formed up and deployed.

Once teams were in place, each group tested their communications and found to their delight that digital VHF communications were possible to both Gainesville and Lake City, allowing them to send radio email messages to anyone out of the Steinhatchee area. This is because digital repeaters can be easily commanded to "connect" and link one to another, much more easily than voice repeaters. Furthermore, HF communications were established by the Bridge communications unit giving another way to send digital email messages.

The Incident Commander now had a means of outside communications should he need it --- and indeed, he was about to need it.

An emergency message over FM Marine Channel 16 was received from the Captain of the cruise ship

PoorNavigator indicating their 19-foot draft had finally grounded solidly about 4 miles SW of the outer marker of the dredged canal out of the Steinhatchee River. Worse, the grounding had significantly damaged onboard systems, so the ship was now without motive power and had only emergency electrical power and was losing potable water pressure. There were significant injuries sustained among the total 600 souls on board, and the ship came to rest in what are normally 10 foot waters, with a list of about 30 degrees port --- a very difficult situation. Starboard life boats were unable to be deployed, and winds and seas --- 10-12 foot waves were making it risky to deploy port life boats. The captain wanted to stay put for now but requested emergency help in evacuating injuries as soon as conditions permitted.

The 0900 ICS 201 Briefing and Incident Action Plan indicated that many boats had sunk in Steinhatchee River, creating submerged and sometimes unseen navigational hazards. Only a fraction of the available rental boats at the three local marinas were still floating and in recoverable condition. There is only one doctor in Steinhatchee, a pediatrician who normally is only available 4 days per week. The nearest hospital is 45 minutes away in normal times, in Perry, Florida. There are two police vehicles stationed in Steinhatchee Florida, and one of them was disabled by a fallen tree. There is one ambulance --- at Fire Station 51 – in Steinhatchee.

These resources aren't exactly what one would wish to evacuate a 600-person cruise ship in a precarious position 10 miles over water from the town, and with multiple injuries among the victims.

Recognizing the significant worsening of his disaster, and the increased resources that would be required, the Incident Commander issued orders to request additional resources from inland:

- Triage medical team & equipment
- 6 total ambulances, split between Type I and Type II
- Water truck
- Helicopter Life Flight & crew on standby for possible Life flight rescues.

The Alachua County ARES volunteers went to work getting these messages out. With non-existent communications to the local counties, they took advantage of the monitoring being provided by Alachua County EOC as well as State of Florida EOC and sent requests to both those organizations (who still had normal communications) by any and all means possible, to allow them to work out the best response. Given the communications difficulties, this complied with the State Comprehensive Emergency Management Plan.

The Incident Commander also ordered Logistics to immediately start locating, acquiring, and staffing up to 20 small vessel rescue boats from the remaining floating stock of the local marinas, recognizing that larger resources were not within 100 miles and would not reach them for many hours at typical watercraft speeds. The three local marinas and many volunteers went into action, locating capable small craft operators, bailing out half-sunk boats, refueling them, equipping them and getting them ready within the protected waters of the Steinhatchee river. With seas still FAR beyond those boat's capabilities, it would be several hours before they would be able to assist in any rescue effort, however.

Exercise Assumptions and Artificialities

Since an exercise isn't reality, there are always assumptions and artificial constraints that are necessary to make practice possible. Exercise participants should make reasonable responses to events as they are presented to gain the most training advantage from the exercise.

Assumptions

Some assumptions are required for this exercise:
1. Participants are here to practice and to learn, mistakes will be made and learned from if we all work together.
2. Although the events may not be perfectly planned, we accept them as plausible and move forward with the exercise.
3. If a real world emergency occurs, it certainly takes precendence over our exercise!

Artificialities

During this exercise, the following constraints:
Exercise communication and coordination is limited to participating exercise organizations, venues, and the systems we're trying to test and practice; refrain from using cell phones!

RULES OF CONDUCT

This is a laid-back training exercise designed to give our group a chance to both practice deployed and long-distance emergency communications, and to immerse themselves in the "paperwork and procedures" of the Incident Command System.

Participants should observe normal driving regulations and safety precautions at all times. We will be operating on the side of one road, and within two commercial restaurants. Certainly position yourself well off the roadway and clearly marked and make no hazard to traffic! Find outside seating if possible at the commercial restaurants (who have been contacted about our exercise) and cause as little disturbance as possible to their normal operations. Have amateur radio brochures available to help the public understand your activities.

SAFETY ISSUES

SHOULD A REAL EMERGENCY OCCUR, USE THE PHRASE "THIS IS A REAL-WORLD EMERGENCY" TO EXPLAIN THAT YOU ARE DEALING WITH AN ACTUAL, RATHER THAN SIMULATED EMERGENCY SITUATION. THESE COMMUNICATIONS HAVE PRIORITY OVER ALL EXERCISE COMMUNICATION.

TRAVEL: This exercise includes significant travel by private vehicle. Check vehicles for fuel, fluids, tires, headlights, brake & turn lights, windshield wipers and other safety equipment prior to departure. Drivers should be aware of the planned route to the Check In point and Staging. Driver should have a paper map to supplement GPS or other navigational aids, as during real incidents, there may be interference to electronic navigational systems. Although group communications may be carried on via cell phone or radio, the driver should remain vigilant and undistracted. Communications duties should be handled by a passenger.

IN THE EVENT OF TORRENTIAL RAIN OR OTHER DRIVING HAZARD WE WILL STOP AT AN APPROPRIATE POINT AND DELAY THE EXERCISE UNTIL DRIVING CONDITIONS ARE SAFE AGAIN.

IN THE EVENT OF A VEHICLE DISABLED, WE WILL ARRANGE FOR NECESSARY REPAIR / TOW / SAFETY AND THE INCIDENT COMMMANDER (JEFF CAPEHART) WILL DIRECT HOW TO FURTHER PROCEED.

CHECK IN AND STAGING: The check-in/staging area is a commercial establishment. Although this isn't a prime season for Steinhatchee, there still could be significant traffic and uninvolved personnel in the area. Proceed slowly and carefully in the check-in/staging area and stay clear of the commercial establishment to avoid interfering with their operations, while attending to parking, restroom usage etc.

DEPLOYMENT LOCATIONS: Steinhatchee is a boating/fishing town. Large boats are frequently towed by ordinary citizens who may have variable levels of towing skill, and have limited visibility of areas around their vehicle/boat. Stay well clear of people towing such objects. Those deployed off the 358 Jena bridge need to be WELL CLEAR of the roadway and well marked with cones, or other objects to reduce risk of injury. Those in parking lots, position yourself at the far end of a parking lot to cause the least interaction with others, put your self between two parked cars with abundant marking. Those in a commercial establishment, obtain your refreshments and position yourselves so as to cause the least intrusion on the normal operations, and use signage to explain to onlookers what is your purpose. Having brochures about Amateur Radio Emergency Communications might be a wise move to help deal with questions while you are busy.

SUN: Susceptible individuals can acquire a painful or dangerous burn in full sun exposure for only minutes. Protect yourself with awnings, canopies, large-brimmed hats, sunscreen, clothing and wise positioning.

HEAT: Florida mid-day temperatures can be brutal and dangerous. If you feel overheated, seek a cooler environment quickly, which might be an airconditioned building or vehicle; get yourself OUT of sun exposure.

CARBON MONOXIDE: be aware of the hazard in any idling vehicle and get fresh air if you develop initial symptoms such as a headache, dizziness, confusion.

WATER HAZARDS: At the water's edge in a saltwater town are SHARP BARNACLES and similar items that can CUT AND INFECT you badly. Avoid entering the water, wear appropriate footwear at all times and be careful of wet/slippery surfaces. If a radio or other item is dropped into the water, it is likely gone forever, so keep cell phones, radios and other equipment firmly under control or safely stowed.

GENERATORS: Check generators for fuel leaks before operation. Do not refuel a hot generator until it has time to cool somewhat.

ELECTRICAL EQUIPMENT: Be very careful of the currents that low-voltage high-power equipment may require. All power circuits must have fuses or circuit breakers and correct sizing of wire. Do not attempt to draw excessive current from a vehicle accessory or cigarette lighter outlet. Be careful when making connections to vehicle electrical systems to properly connect, avoid reverse polarity, and avoid any kind of dead short across vehicle high current supplies.

ELECTROCUTION: Do not allow water to reach extension cords and connections.

MICROWAVE: Although our microwave equipment is relatively LOW POWER, do not allow directional antennas to be pointed at nearby animal or human life for even seconds. Keep equipment unpowered until it is securely positioned.

ANTENNA PLACEMENT: Be cautious when emplacing antennas. Do not attach anything to railings. Use appropriately secure bases for VHF or HF or microwave antennas.

LOGISTICS

1. Participants will meet for breakfast at the County Foodly restaurant (or parking lot) by 0745. They should bring go-box communications gear and other equipment suitable for their assignments, and also suitable clothing depending on their outdoor/indoor assignment. Vehicles for travel to/from Steinhatchee will have been pre-arranged, and should be in good working order with adequate gas, oil, safe tires, etc.
2. Communications between vehicles on the way out will be by Simplex FM on 146.52 MHz.
3. Drivers/or team leaders will also exchange cell phone numbers in case someone gets lost or has a significant vehicle difficulty.
4. **DESTINATION: CASEY's COVE: 4527 SW Highway 358, Steinhatchee, FL 32359-8116** We'll travel south on 34th street to 39th avenue, then turn left when we reach CR-241, and proceed south; then east on Newberry Road, until we merge into US 19. We can make a bathroom stop at a Hardees on the way. We'll turn left onto FL358 about 10 miles west of Cross City, proceed to the check-in location at Casey's Cove; park out of the way. Another bathroom stop is possible there. The Planning Section is encouraged to have a small portable table for this purpose, possibly with a canopy.
5. Complete your checkin procedures there and proceed with the Exercise.
6. After the Exercise is concluded, your Team will proceed to Roy's Restaurant for a hotwash

debriefing section and all demobilization paperwork, which can be completed either in the Roys Restaurant parking lot or in their (small) lobby and gift shop. The Planning Section is encouraged to have a small portable table for this purpose, possibly with a canopy for outside deployment out of the way of normal restaurant traffic. Do not set tables etc up inside of Roys!

7. After lunch, the Planning Section will ensure that all Participants are accounted for and transported back out of Steinhatchee.

8. FM Simplex 146.52 will be a communications tool on the trip back.

SECURITY

There is no specific "security" for this exercise. Please keep watch over your communications gear / generators /etc. Lock your vehicle when you leave it with valuable gear inside of it. Utilize chains or other protective measures to protect valuable gear such as a generator in your open truck bed.

COMMUNICATIONS

1. During transport to and from Steinhatchee, both cell phones and FM Simplex 146.55 will provide communications.

2. During the actual Exercise, please discontinue cell phones unless you have a situation you cannot resolve over the assigned radio frequencies, or there is a safety issue or security issue that must be addressed.

3. Please be certain that your cell phone and radio gear batteries are well charged prior to the beginning. You may wish to bring vehicle charging equipment.

4. During the exercise, at least 4 location will be staffed with varying bands --- please see the ICS-205a ADDENDUM for detailed information on HF, VHF, Microwave frequencies, emissions, callsigns, modes, as well as IP numbers on relevant Ethernet local area networks for web server, voice over IP telephones etc., that are involved in this test. To avoid having multiple sources of information that may not remain "in-sync" that information is present on the ICS205A ADDENDUM.

5. **SHOULD A REAL EMERGENCY OCCUR, USE THE PHRASE "THIS IS A REAL-WORLD EMERGENCY" TO EXPLAIN THAT YOU ARE DEALING WITH AN ACTUAL, RATHER THAN SIMULATED EMERGENCY SITUATION. THESE COMMUNICATIONS HAVE PRIORITY OVER ALL EXERCISE COMMUNICATION.**

SCHEDULES

0645   Earliest that you might want to arrive at Country Foodly

0800   Form up in the parking lot for the trip to Steinhatchee. Planning Section Chief is in charge of keeping manifests for transportation.

0930   Approximate arrival time to Steinhatchee (it is 65 miles, with several slower points. Drive to CASEY's COVE for check in.

1030   Approximate time of checkin & deployment of teams

1045   Approximate time of teams reaching operational capacity. The Incident Commander will be keeping track and will provide the 1000 Briefing updates on Assignments, allowing you to commence with long distance formal traffic assigned to your position.

1145   Approximate time that formal communications will likely be completed and that the Incident Commander will authorize you to cease communications operations and begin to move to Roy's Restaurant

1200   Approximate time that demobilization paperwork will be in process at Roys Restaurant and participants will move inside to obtain seating for their hotwash debrief. We will arrange ahead of time with Roy's

1330   We expect to depart from Steinhatchee by this time.

1500   We expect to return to Gainesville / Country Foodly by this time. Please keep the Planning Section updated on your disbursement of participants.

MAPS AND DIRECTIONS

Addresses:

County Foodly:	5240 NW 34th Blvd, Gainesville, FL 32605
	352 377 7863 Breakfast
Hardee's Old Town:	Hwy 351-U, Old Town, FL 32680
	352 542 8887 Bathroom stop
CASEY's COVE:	4527 SW Highway 358, Steinhatchee, FL 32359-8116
	352 498 1061 Check-in Station
Good Times Marina:	7022 SW 358 Hwy, Steinhatchee, FL 32359
	352 498 8088 **COMM UNIT #1**
Hungry Howies:	806 S. Riverside Dr., Steinhatchee, FL 32359
	352 498 7100 **COMM UNIT #2**
BRIDGE	FL-358 crossing the Steinhatchee River.
	COMM UNIT #3
JONESBORO LOOKOUT	Intersection of FL358 and US-19 **COMM UNIT #4**
ROY's RESTAURANT:	100 1st Ave S, Steinhatchee, FL 32359
	352 498 5000 LUNCH & Hotwash Discussion

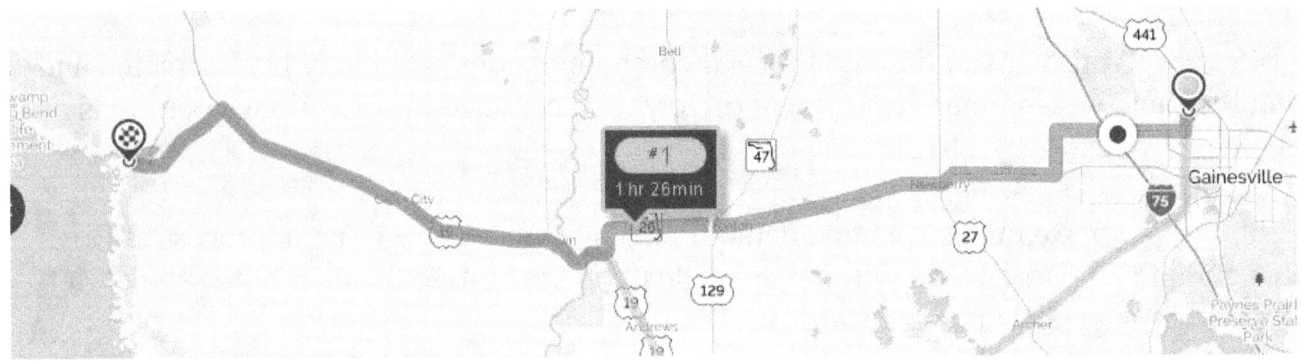

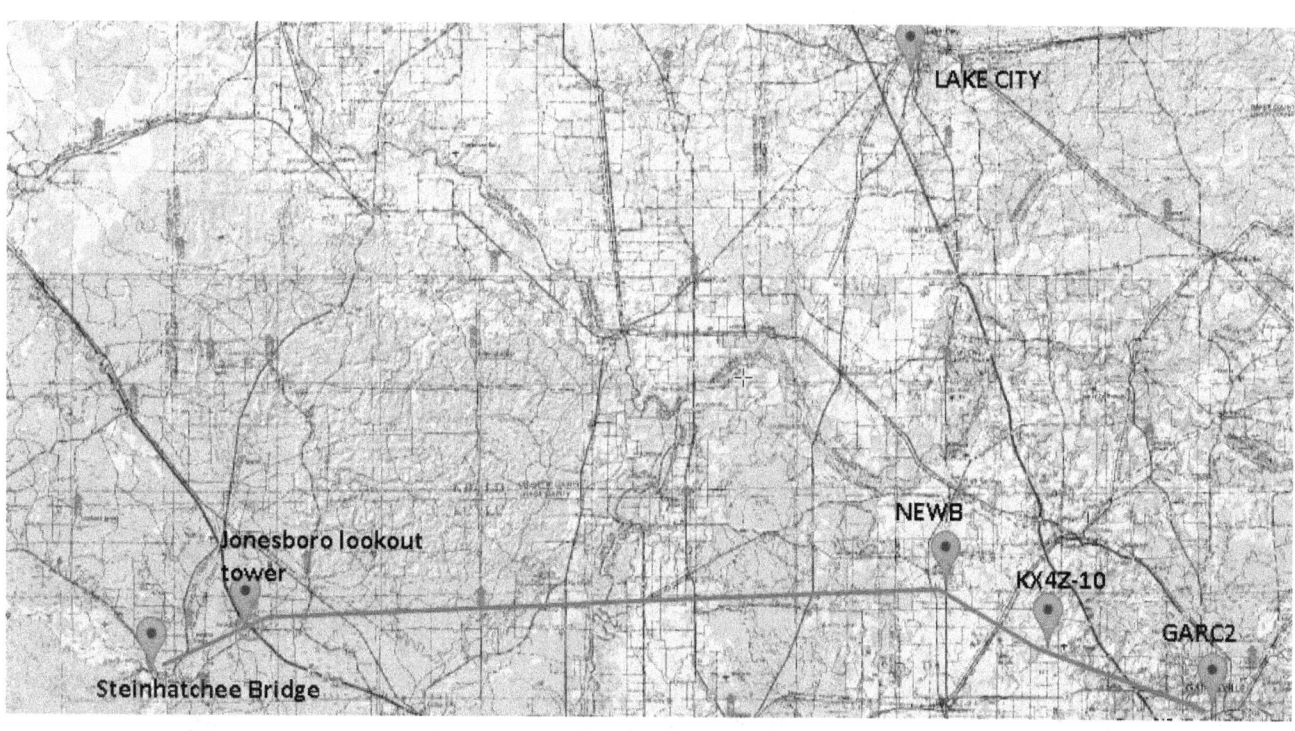

SECTION 3: THE NUTS & BOLTS OF PLANNING

PERMISSIONS

Our group had previously learned how to get permission for the use of Florida fire look out towers --- identify the tower on the free maps provided at https://tnlandforms.us/fltowers.html   Then find which Division of the Florida Forest Service (http://www.freshfromflorida.com/Divisions-Offices/Florida-Forest-Service) controls that area, and call the official in charge to find out if the tower can be utilized. The *tnlandforms* map is often out of date and towers may not exist or may have been sold. Using satellite maps you may be able to discover if the tower is still in existence; however one exciting road trip for me (during which I got completely stuck in soft sand) resulted in discovering that a particular tower was inaccessible due to private land and locked gates....had I asked the Florida Forest Service officials first, I might have avoided that exciting trip.

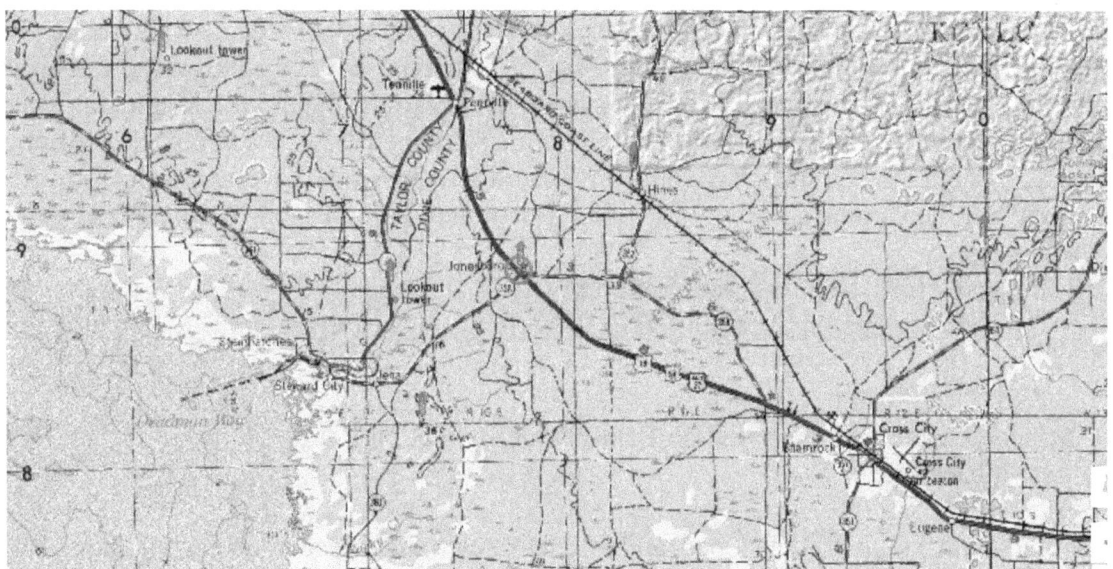

Topo map from https://tnlandforms.us/fltowers.html with our chosen lookout tower circled

In our case, the tower we wished to use had been sold, so gaining permission became more difficult. Even finding the current owner was difficult and required using the county property appraiser's web site and maps to identify the parcel, and who was paying the taxes for the land on which the tower sits; then a phone call expedition to try and reach the owner.   Tall towers that are almost 100-years old have significant liability associated with them, so the owner was quite willing for us to use the tower --- if we obtained liability insurance. Trying to find liability insurance for such a unique structure is diffcult and my homeowners liability policy company would not touch it. However, I subsequently found an Event Planner company that even after a careful explanation of exactly what we needed, was willing to insure us for 1 day for a reasonable amount. (Check out: https://www.theeventhelper.com/)

MARINA / Restaurants: As a result of my reading to write the scenarios and plans, I was well aware of the disastrous flooding in 2016 of the very town to which we were going to deploy. Thus, in contacting the locations we needed for our teams, and the restaurant where we wanted to hold our "hotwash" luncheon, I was able to explain how our volunteer effort related to real events they had endured just the previous year. We got a very warm reception from all the groups from whom we needed permission. I subsequently mailed a copy of our Exercise plan to the major locations we would be using. Because we

SHERIFF: Because we needed to park a truck near a bridge off the roadway, I felt it would be important to notify the local authorities. You speak to the dispatcher (don't call their 911 number!) who gives a message to a busy deputy who may or may not call back. The dispatcher was very enthusiastic about our plans; the deputy didn't call back until my third try, but thanked me for calling --- and crucially gave me his name/telephone number to have anyone call if we had any issues! Huge help!!

TRAINING

1. I had to build two microwave-based systems; most of the other gear we already had built from our previous exercises. Everything has to run from a battery or a vehicle, and microwave systems require various voltages such as +5, +12 etc to run the network switch, the Grandstream VOIP phone (+5V) and the Ubiquity device (+15-+24 "power over ethernet"). I built the stations on plywood for simplicity, one on a flat board for ease of construction (that system went into the pickup truck) and one system on a small open plywood box with carry handles to go into the restaurant. Each was powered differently: the truck-based system turned out to run well from a modified sine-wave inverter output, powering all the "wall warts" for the various devices. The other system had to run from a 7Ahr 12VDC gell cell ---

20VDC achieved with a step up dc-dc converter:
(https://www.amazon.com/gp/product/B00ID3TJ3U/ref=oh_aui_search_detailpage?ie=UTF8&psc=1

5 VDC achieved using a cigarette-lighter USB charger (break into, solder wires to the +12 V input system, re-glue it back together)

With the microwave systems developed, and AREDN software uploaded on each system, static IP numbers were assigned on each system so that the computer and voip phone would have known addresses.

WINLINK TELNET:
The unit that would be deployed to one of the restaurants would use WINLINK EXPRESS using TELNET to send/receive messages by tcp/ip microwave to the BRIDGE station – the TELNET setup

on WINLINK express offers an option to connect not over the internet to a CMS (Central Message Server) but instead to a specific RMS_RELAY instance. We used this option, inserting the static ip address of our portable Gateway. The Microwave Ubiquiti devices were key to all of this, because they internally run a tiny linux computer that provides DHCP services; for both the deployed station and the truck relay station, it was the microwave systems that provided the equivalent services of a home wifi router, determining the IP numbers for all the connected devices.

For the truck-mounted RMS_RELAY system there would be multiple devices on the network:

Ubiquity Bullet transceiver	Flat Board Microwave with Ubiquiti Bullet NF4RC-003　WIFI: 10.254.254.3　LAN side: 10.247.240.25　(http: on 8080)	provided microwave connection and DHCP services
Lenovo computer running WINLINK RMS_RELAY and RMS_TRIMODE	10.247.240.29　(RMS_RELAY on 8772)	Provides temporary storage of emails and coordinates HF sending and receiving of email via an ICOM 728 transceiver
Raspberry Pi linbpq system	NF4RC-3　10.247.240.27	Provides VHF port for access to the RMS_RELAY over VHF; also provides node switching allowing connection over AX.25 packet either way between the mcirowave network and the VHF AX.25 system on 145.070
Grandstream voice over IP phone	10.247.240.26	Provided voice connections from the Bridge to Comm Unit #2

With most or all of this complicated microwave system constructed, it was demonstrated at two different monthly training meetings, to better acquaint our volunteers with dialing by IP and using voice over IP. However, we didn't have the connections between the Raspberry and the microwave working properly by the time these training meetings were scheduled.

TABLETOP DRILL

We setup all 4 stations during the Tabletop drill --- and this was quite an undertaking because this occurred almost immediately after our group had for-real staffed up to 9 shelters and the EOC during Hurricane IRMA. Although significant damage was feared, peak winds didn't exceed 61 mph so our County was largely spared, and I thought we should go ahead with the Exercise and the training for it.

After Action Report / Improvement Plan (AAR/IP)   Alachua County ARES   2017 Hurricane Test

There were lots of issues during the TableTop Drill --- primarily, the RMS_RELAY software was incorrect configured

A list of all the issues discovered at the TableTop is as follows:

No.	Issue	Resolution
1	Contrived scenario	Best I could do. Mexico City just had a 7.1 Earthquake which collapsed > 40 buildings, destroying a lot of communications infrastructure. We could pretend we were in Mexico and have the exact same communications needs. Exercise scenarios are always somewhat less that perfect because we practice things that RARELY HAPPEN....just like militaries prepare for nuclear war....which has NEVER happened....yet. Puerto Rico currently has nearly ZERO utility power for the entire island --- never happened before. Strange things happen. **Improvement:** Sending WEATHER BULLETINS to the Incident Commander may be an important thing to retrieve; may be able to use CATALOG features of WINLINK for this. Improvement: Puerto Rico is sending out Health and Welfare speadsheet info to Red Cross...maybe we should do this ?
2	Defective PACTOR modems	**RESOLVED.** New bluetooth modem purchased –pactor III. By being BLUETOOTH, yet another chance for crashing a USB card is removed.
3	NF4RC emergency WINLINK gateway was not accomplishing HF forwarding – and actively refusing connections.	**Believed resolved.** Testing continues. This was several problems, took > 6 hours to resolve. 1. NF4RC's setup was derived from a previous temporary server originally configured for KX4Z. The call sign was correctly changed to NF4RC in the top level configurations of both RMS_RELAY and RMS_TRIMODE.....but not in the frequency assignments of RMS_TRIMODE. This error causes the system not to display on WINLINK MAPS, or exist in WINLINK Channel lists.

		2. NF4RC's setup was accomplished, in part, while out of connection with the Internet (both due to spotty internet access post IRMA, and due to an erroneous understanding of setup requirements). As a result, a couple of CHANNEL FORWARDING screens failed to load with available RMS's for forwarding options [Note: it turns out that at times a specific sequence of moves can be necessary to accomplish this.] 3. A crucial error was setting an "Extra Time To Forward" setting to 0 minutes by accident, instead of the default 10 minutes. 4. Firewalls on one or both computers tested were improperly configured and refused incoming tcp/ip connections to RMS_RELAY; this made it impossible for COMM UNIT 2 to access using WINLINK TELNET. 5. **CMS FORWARDING was not properly configured. All checkboxes on the HF Forwarding to CMS need to be checked!**
4	KX4Z-7 was having problems connecting to NEWB	**RESOLVED.** The reason was that NEWB was dying with a dead marine 12V battery; the reason for that was that electrical wiring in the panel box at the base of the Forestry Lookout Tower was loose in the screws of the GFCI breaker in the panel box. Once this was discovered, the screws were tightened, and 120VAC again was available to run the battery charger / battery maintainer and NEWB came back alive.
5	Packet Storm	**Ameliorated.** Possibly because of the dying node NEWB, there was a huge amount of repetitive packets in an attempt to get packets through. I had asked TWO users to make a complicated and lengthy set of VHF packet connections: Station → Bridge → Tower → NEWB → KX4Z-10 Making FOUR packet relays in and of itself causes a lot of packets to fly about, but having two stations attempting it simultaneously – and then having NEWB dying and requiring multiple multiple repeats as its signal disappeared (the 12V

| | | battery was down to 9 volts; the power output from NEWB was down to 3 watts) caused complete packet storm on the channel, with every available time slot taken.

The solution to this was to drop NEWB out of the loop, and to have one station at a time try to make their required connections.

The LEARNING POINT is that **listening** to the packet channel can warn you when the channel is being overloaded. Unlike analog voice duplex repeaters, packet channels CAN handle more than one user "simultaneously" --- but once you crowd the channel, the throughput drops DRAMATICALLY…and in our case, to zero.

For the S.E.T:
1. COMM UNIT 1 has a 40 watt transceiver – and thus probably can skip the BRIDGE and make their connections much simpler:

AC4QZ → TOWER → NEWB → KX4Z-10/7

2. COMM UNIT 2, if they utilize VHF packet by gaining telnet access (PuTTY) to NF4RC-3….should LISTEN to the channel and avoid overloading the channel if COMM UNIT 1 is trying to get a message across. Gentle sharing of the channel should work much better this way, and COMM UNIT 2 can still make keyboard to keyboard connections

COMM UNIT 2 – PuTTY to NF4RC-3
NF4RC-3 → TOWER → NEWBERRY → proxy station in Gainesville

Huge learning point: every station that is using a packet channel would do well to either listen to the channel or to watch their soundmodem or other display of all channel traffic so as to avoid packet storm overloading of the channel. Packet can handle SOME simultaneous users (unlike analog voice) but not a LOT of simultaneous users. |
|---|---|---|
| 6 | Unfamiliarity with ICS forms; some General Staff didn't have forms for their functions | **RESOLVING:** We are all new at these and learning all at the same time. I think each General Staff member knows which forms are "theirs" now. A new printer has been procured for the Safety Officer who was having problems printing things. |
| 7 | COMM UNIT 2 go-box was not fully assembled. | **RESOLVED**: Go-Box is now fully assembled and delivered to Susan Halbert. Battery needs overnight charging to be fully |

		charged.
8	12V power from the truck is not yet arranged. Need to add connections directly to the battery system as the power required exceeds what should be drawn from the cig. lighter.	**RESOLVED.** Powerpole as well as cig lighter socket attached to main battery wiring through a 25A fuse. . Power pole "extension cord" also built.
9	Contact with the local sheriff has been attempted on two occasions but the deputy has never called back.	RESOLVED: contact made and appreciated.
10	Need to discuss with local press	**RESOLVED:** Press releases sent to 5 different outlets.
11	Need brochures about ham radio	**RESOLVED:** G. Gibby found some on the ARRL web site and ordered them
12	Need some sort of POSTER for each station	**RESOLVED** I got in ARRL handouts and I made color 8.5x11 placards that can go in sheet protectors and be affixed to each table.
13	Try and get new people interested in ham radio for the Dixie County Amateur Radio Club	Dixie County leadership was unable to join us
14	Proxy for the COAST GUARD	RESOLVED: Ohio ARES group is willing to do this.
15	Proxy for the EOC	Resolved: Nancy filled in.
16	BADGES There were plans to take photos of members – but these don't seem to have happened.	RESOLVED: Art came up with "Participant" badges that worked out very well.
17	EOC Involvement John Shaw indicated we would have EOC involvement but I haven't made any contact since except for getting Jeff B. to answer several questions for me.	The EOC was not able to be involved.
18	Difficulty remembering IP numbers	**RESOLVED** – IP numbers are now pasted on both the COMM UNIT 3 board and the COMM UNIT 2 go-box
19	Limited access to ethernet from ACER computers....have only 2 usb-adapters; and multiple people were connecting on COMM UNIT 2's microwave connection (which is good example of unexpected increased learning)	RESOLVED: purchased 2 more USB toether adapters

20	Identifying Automobiles as "official"	**RESOLVED**: Extra "AMATEUR RADIO COMMUNICATIONS" magnetic signs were purchased. *REMEMBER they CANNOT be on the car at highway speeds….keep them inside and post them only when arriving at location.*
21	Caution Tape for Bridge Unit	**RESOLVED:** Caution tape purchased
22	Support system for BRIDGE microwave	**RESOLVED:** 12 pound tripod system purchased (Amazon) to hold the microwave system
23	Securing Permission	**RESOLVED** Permission has been secured for all facilities. Next need to send confirmatory mail to each facility.

SECTION 4: AFTER ACTION REVIEW: EXERCISE OVERVIEW

Exercise Details

Exercise Name
2017 ARES Steinhatchee Storm (Simulated Emergency Test)

Type of Exercise
Full Scale Exercise

Exercise Start Date
October 7, 2017

Exercise End Date
October 7, 2017

Duration
6 Hours

Location
4 locations: Steinhatchee Good Times Motel & Marina; Steinhatchee Hungry Howies; Bridge over Steinhatchee River and the Jonesboro Lookout Tower near Steinhatchee.--all 60 miles west of our County in a on-the-road deployment.

Sponsor
Alachua County ARES, a component of the American Radio Relay League (ARRL)

Program
Amateur Radio Emergency Service

Mission
Communications Support

Capabilities
VHF local communications, analog voice and digital (AX.25 packet)

HF local and national communications, digital WINLINK.

Radio Email to anywhere, via WINLINK

Microwave digital communications including Voice Over Internet Protocol (VOIP) and WINLINK telnet.

Scenario Type

Total Communications Failures

Exercise Planning Team

Gordon L. Gibby MD KX4Z NCS521

Jeff Capehart W4UFL

Participating Organizations

Alachua County, Florida

Alachua County ARES

State

Volunteers from Sarasota, N4SER

Volunteers from Marion County Hospital Emergency Communications

Volunteers from Ohio Black Swan.

Number of Participants
- Players - 11
- Controllers - 0
- Evaluators – 0

Amateur Radio Volunteers at the Gainesville Senior Center (a hurricane shelter)

ARES volunteer conducting voice and digital VHF communications from Good Times Motel & Marina

SECTION 5: AFTER ACTION REPORT
EXERCISE DESIGN SUMMARY

Exercise Purpose and Design

Although, this exercise was designed well before Hurricanes IRMA and MARIA devastated Puerto Rico, the skills we sought to develop, exercise and test are exactly those which 50 amateur radio volunteers for the American Red Cross are using to help Puerto Rico.

These same systems are actually in daily use by some of our volunteers, and one of our volunteer's stations is being used to assist as a WINLINK email relay station for Puerto Rico. So this exercise is very relevant to modern amateur radio emergency communications.

This exercise followed our May 2017 Hurricane Test, which was our first full scale exercise built on the Incident Command System procedures. In this exercise we wanted to further our training by

- deploying to a distant location
- setting up without any commercial power
- providing communications, both voice and digital back to the "uninvolved" Alachua County
- testing both HF and VHF long-distance communications
- testing microwave amateur radio communications, both digital and VOIP

Training for this type of scenario – total loss or overwhelming of local conventional communications systems (telephone/Internet) began 18 months before this exercise, as local skills, assets, and strategies began to be sharpened.

Development of this exercise began months before, with testing of communications from the Steinhatchee area, and the initial testing of the Ubiquiti microwave systems.

An Exercise Plan was created with full details for participants. All relevant ICS forms were created.

Exercise Objectives

- **Objective 1:** Assess the capabilities of our group to work within the Incident Command System framework on a deployed mission outside of Alachua County.
  Capability: ICS FORMS

- **Objective 2:** Assess the capability of our group to provide a 60+ mile digital VHF communications link, and provide practice to our members in making multiple connections to reach a distant station.
  Capability: ANTENNA PLACEMENT, MOBILE DEPLOYMENT

- **Objective 3:** Provide practice for, and assess our capabilities at sending/receiving WINLINK email messages and attachments by VHF and/or HF and/or Microwave technologies.
  CAPABILITY: ANTENNA PLACEMENT, WINLINK COMMUNICATIONS, BACKUP POWER, MOBILE DEPLOYMENT

Scenario Summary

October 7, 2017

Hurricane Thoughtless has been moving through the Gulf of Mexico for 2 days and finally settled in on a course that took it right to Steinhatchee, Florida, a sleepy fishing, tourist-scalloping town of about 1200 residents that has been flooded many times by hurricanes.

As the projected path of the Hurricane became more confident, about half the population of Steinhatchee voluntarily evacuated inland, remembering the recent flooding brought about by Hurricane Hermine that shut down large parts of the town for weeks and months. The Sheriff of Taylor County recognized the developing situation and activated the ICS system, becoming the Incident Commander, with careful communications to the Sheriff of Dixie County (beginning on the southern edge of the Steinhatchee River) as well as the town leadership of Steinhatchee, Keaton Beach (11 miles to the north) and Horseshoe Beach (to the south).

The Incident Commander took preparatory steps to have citizens prepare for the oncoming hurricane in the evening hours of October 6th, as winds were rising. Water was stored as much as possible, debris secured, boats secured as much as possible, and everyone battened down the hatches for the coming storm, which was strengthening into a Category II hurricane.

Unbeknownst to the Incident Commander, cruise ship PoorNavigator was being driven by the storm and moving in the direction of their area, driven with the storm surge that allowed its 19-foot draft to enter the relatively shallow waters of the northern Gulf of Mexico. A bit behind the curve, the Captain of the cruise ship did not recognize the full implications of his loss of control of his track; similar to the tragic course of the barge El Faro that intentionally held course right through a hurricane a few years back, with the sinking and loss of all the crew.

As the storm strengthened, the Incident Commander grew increasingly concerned about the possible damages, which might be greater than what had been experienced by Steinhatchee in 2016. When the storm made landfall at 0400, not only did all electrical power go out (quite as expected) but also within 30 minutes, normal telecommunications, including cell phone, landline, and Internet also quit. Communications were still possible for another half hour with the remotely operated Coast Guard Marine Radio tower in Horseshoe Beach, but by 0500 internet control of that radio failed and the Incident Commander realized that his communications extended only as far as his marine FM transceiver could reach --- not a happy feeling. He decided to send a courier inland to try and ask for assistance. At 0600 a courier was able to leave the area, making a circuitous route to avoid the peak intensity of the inland hurricane, and found working telephone service at Old Town, Florida, and contacted State of Florida Emergency Management in Tallahassee. A request for some form of "longer-distance communications support" was communicated since normally Steinhatchee residents simply endure hurricanes to the best of their ability, but the Incident Commander did want the ability to request additional resources quickly should something unexpected happen.

The State EOC was watching the storm now turn toward Tallahassee and quite concerned about communications losses in the capital city that would need large amounts of cellphone capacity, but he was aware of a volunteer communications ARES group in Alachua County and reached out to the Alachua County EOC to see if they would be able to reach the area before larger satellite trucks could move in from caches throughout the Southeast. The Alachua County emergency manager contacted Jeff Capehart of the Alachua County ARES group (a part of ESF#2) and Jeff's response was eager and enthusiastic. As winds were really not dangerous in Gainesville, he knew a good group of communicators would be having their traditional Saturday breakfast at 0730. He thought he could muster a significant set of resources by 0830 or earlier to head westward, as the storm was weakening and sharply turned toward Tallahassee.

Realizing that the Alachua County ARES group has specific digital capabilities involving email, the Alachua County EOC agreed to monitor a prescribed email address for communications, in addition to remaining available on the 146.82 repeater and 146.52 simplex. The State EOC also indicated that they would also monitor a prescribed email address, but that they currently did not have packet amateur radio access.

Thus, by 0830 a caravan of volunteers headed westward from Gainesville toward the Check-In station at Casey's Cove just outside Steinhatchee, with multiple VHF, HF and even microwave gear. Permission to use the antenna on top of the Jonesboro lookout tower had been graciously provided by its private owner, who had purchased the tower years ago from the Florida Forest Service. The team thought that not only were HF connections quite easily made with portable stations, but VHF digital packet communications could be achieved by placing a portable "node" station at the Jonesboro lookout tower, and another portable "node" station near the highest point in town --- the bridge over the Steinhatchee River.

Upon reaching the Casey's Cove check- in station (a gas station/deli/convenience store with a large parking lot) the Incident Commander at the Good Times Marina bar and grill ("WATS DAT") was notified by FM marine radio.

Participants will fill out the ICS – 211 Incident Check-In List

Teams were given copies of the 0900 ICS 201 Incident Briefing, and of the 0900 Incident Action Plan, including multiple forms. Instructions (ICS form 204) were quickly given to emplace units at

- Incident Command Post, Good Times Marina
- Logistics station at Hungry Howies
- high ground at the FL-358 bridge over the Steinhatchee River
- Jonesboro lookout tower at the intersection of FL-358 and US-19

and the Communications Teams quickly formed up and deployed.

The Incident Commander now had a means of outside communications should he need it --- and indeed, he was about to need it.

An emergency message over FM Marine Channel 16 was received from the Captain of the cruise ship PoorNavigator indicating their 19-foot draft had finally grounded solidly about 4 miles SW of the outer marker of the dredged canal out of the Steinhatchee River. Worse, the grounding had significantly damaged onboard systems, so the ship was now without motive power and had only emergency electrical power and was losing potable water pressure. There were significant injuries sustained among the total 600 souls on board, and the ship came to rest in what are normally 10 foot waters, with a list of about 30 degrees port --- a very difficult situation. Starboard life boats were unable to be deployed, and winds and seas --- 10-12 foot waves were making it risky to deploy port life boats. The captain wanted to stay put for now but requested emergency help in evacuating injuries as soon as conditions permitted.

The 0900 ICS 201 Briefing and Incident Action Plan indicated that many boats had sunk in Steinhatchee River, creating submerged and sometimes unseen navigational hazards. Only a fraction of the available rental boats at the three local marinas were still floating and in recoverable condition. There is only one doctor in Steinhatchee, a pediatrician who normally is only available 4 days per week. The nearest hospital is 45 minutes away in normal times, in Perry, Florida. There are two police vehicles stationed in Steinhatchee Florida, and one of them was disabled by a fallen tree. There is one ambulance --- at Fire Station 51 – in Steinhatchee.

These resources aren't exactly what one would wish to evacuate a 600-person cruise ship in a precarious position 10 miles over water from the town, and with multiple injuries among the victim

Recognizing the significant worsening of his disaster, and the increased resources that would be required, the Incident Commander issued orders to request additional resources from inland:

- Triage medical team & equipment
- 6 total ambulances, split between Type I and Type II
- Water truck
- Helicopter Life Flight & crew on standby for possible Life flight rescues.

The Alachua County ARES volunteers went to work getting these messages out. With non-existent communications to the local counties, they took advantage of the monitoring being provided by Alachua County EOC as well as State of Florida EOC and sent requests to both those organizations (who still had normal communications) by any and all means possible, to allow them to work out the best response. Given the communications difficulties, this complied with the State Comprehensive Emergency Management Plan.

The Incident Commander also ordered Logistics to immediately start locating, acquiring, and staffing up to 20 small vessel rescue boats from the remaining floating stock of the local marinas, recognizing that larger resources were not within 100 miles and would not reach them for many hours at typical watercraft speeds. The three local marinas and many volunteers went into action, locating capable small craft operators, bailing out half-sunk boats, refueling them, equipping them and getting them ready within the protected waters of the Steinhatchee river. With seas still FAR beyond those boat's capabilities, it would be several hours before they would be able to assist in any rescue effort, however.

Some of the Comm Unit #4 volunteers getting a "consult" on reducing transmission line losses.

SECTION 6: AFTER ACTION REPORT ANALYSIS OF CAPABILITIES

This section of the report reviews the performance of the exercised capabilities, activities, and tasks. In this section, observations are organized by capability and associated activities. The capabilities linked to the exercise objectives of Operation Hurricane Test are listed below, followed by corresponding activities. Each activity is followed by related observations, which include references, analysis, and recommendations.

CAPABILITY 1: ANTENNA PLACEMENT

Capability Summary: Fixed, pre-existing antennas should obviously perform for necessary communications. However, these may be damaged by high winds, and volunteers need to have the skills to efficiently replace them with ad-hoc created or installed antennas.

HF Dipole Antenna between trees at Comm Unit #4

Activity 1.1: Except for COMM UNIT #4 at the Jonesboro Lookout Tower, every unit had to establish their own antennas – particularly the Bridge COMM UNIT #3, which had to establish HF, VHF and Microwave antennas.

Observation 1.1.1: STRENGTH. COMM UNIT#1 utilized a mag mount VHF portable antenna with great success. COMM UNIT#2 used the built-in antennas of their microwave unit. COMM UNIT#3 put up a random-length HF dipole within 45 minutes using slingshots in dense brush leading to marsh, utilized a mag mount VHF antenna, and mounted an omnidirectional microwave 6 dBi antenna on a 6 foot realtor's sign tripod. COMM UNIT #4 moved a VHF voice antenna 40 feet up the lookout tower. Antennas were a strength of our volunteer group in this exercise and were all up within the first hour.

The estimated setup time until "READY TO OPERATE" for each of the Communications Units were as follows:
COMM 1: 30-40 minutes
COMM 2: 10 minutes
COMM 3: 1 hr for antennas, but then 2 hours for software replacement
COMM 4: 15 minutes

Note: When Comm Unit #4 arrived at the Lookout Tower, the first thing that happened was a vagrant climbed down from the tower and left the scene!

Analysis: The work we have done in building our own portable antennas and developing skills at antenna placement are paying off.

Recommendations:

- **During this exercise, antennas were a strength. Continue the skill training and purchase additional VHF magnetic mount antennas as existing stock degrade.**

CAPABILITY 2: EMERGENCY SIMPLEX REPEATER

Capability Summary: If existing duplex amateur voice repeaters are overwhelmed, or out of service, a portable simplex repeater (that acts like a digital voice recorder, and replays over the air, from a high location, messages received) can provide voice coverage to a devastated area.

This capability was not tested during this Full Scale Exercise

CAPABILITY 3: WINLINK COMMUNICATIONS

Capability Summary: WINLINK provides a world-wide, radio-based email capability that has been leveraged by mariners, emergency communications personnel, missionaries, and the Federal Government. Allowing both email and attachments, it can speed digital messages toward areas where the Internet is still working, and then forward them by the far-faster internet email facilities, or in a complete national disaster, can slowly move email to "Message Pickup Stations" by radio alone. It is the premier HF radio-based email system in the world today.

> **Activity 3.1** Generate, forward, and retrieve multiple emails and attachments via WINLINK, either using HF VHF or microwave capabilities.
>
> > **Observation 3.1.1: MIXED** Our volunteers were very capable at WINLINK communications during this Exercise --- COMM UNITS #1, #2 and #3 were vigorously sending and receiving WINLINK messages. This was happening on three different frequency bands --- HF, VHF, and even microwave. A comment at the hotwash session was that the scheduled meetings we have been having have made for a lot more practice at these skills (as well as on-the-air scheduled events).
> >
> > However, a **WEAKNESS** was observed in the SOFTWARE of the WINLINK system, and in our preparations to deal with such difficulties: The crucial RMS_RELAY/RMS_TRIMODE server system at COMM UNIT #3 failed to initialize properly at the beginning of the exercise. (Appendix A, Item #1) Repeated efforts to restart were unsuccessful. A WEAKNESS was that complete installation copies of all the software had not been brought to the Exercise. Attempts to install an older version of RMS_TRIMODE brought the system up, but then connections resulted in "invalid password" failures.....very frustrating. Further, the older setup did not support automated forwarding of messages, so manual forwarding was attempted with limited success.
> >
> > Finally, a complete installation set of software was downloaded over the Internet (using a cell phone as a hot-spot) and installed – which then proceeded to work admirably. However, this was after more than 2 hours of work, putting the Exercise severely out of schedule, and the propagation to the pre-selected Target station (N5TW) was now marginal, resulting in the usage of other stations, particularly around northern Georgia.
> >
> > > **Analysis** Our local ARES group has developed a significant strength in this area, but we have not yet prepared adequately for failures of the system itself.

Recommendations:

- **Continue WINLINK practice.**
- **ALWAYS equip every comm unit with complete re-installation package on a USB flash drive.**
- **Train on peer-to-peer WINLINK skills.**
- **Work toward having alternatives to WINDOWS-based software.**

CAPABILITY 4: BACKUP POWER

Capability Summary:

Electrical Utility power loss is one of the most frequent occurrences in hurricanes, and is a major cause of loss of traditional communications. Amateur radio emergency volunteers need to have alternate power capabilities.

> **Activity 4.1** ALL stations active in this Exercise operated out of battery or vehicle power throughout the entire exercise.
>
>> **Observation 4.1.1: Strength.** Simply put, it worked. Electrical backup power worked well throughout the exercise at all units.
>>
>>> **Analysis** Considerable effort into this strength has borne results. It was practiced at the table-top exercise conducted just a few weeks before the Full Scale Exercise; power connections directly to the battery system of the pickup truck for the Server were established and worked well, though it would be better to arrange for wiring that could be maintained during vehicle movement. A microwave station was built with battery power, and it worked well for this exercise.

Recommendation:

- **Continue to develop strengths.**
- **Develop high amperage pickup truck power that can be maintained in motion.**

CAPABILITY 5: MOBILE DEPLOYMENT

Capability Summary: In a true communications emergency, it is likely that there will be additional locations that suddenly develop an emergency need for communications. Amateur radio volunteers should maintain the ability to service those needs through mobile vehicles, possibly including dis-mountable VHF and HF gear that can be set up quickly at a new fixed site, including antennas.

Activity 5.1 Our entire team traveled 60 miles to our simulation environment.

> **Observation 5.1.1: Strength.** Our entire team deployed without incident and maintained VHF simplex communications on 146.55 MHz VHF FM throughout the move; some stations corrected settings on their radios to make this succeed. One vehicle installed mobile equipment for this exercise but seemed to have an intermittent mic connection.
>
> **Analysis** Demonstrated the growing capabilities of our group – I am not aware of any similar 60 mile deployment of any other ARES group in Florida during the S.E.T.

Recommendation:

- **Encourage members to continue to obtain and install mobile vehicular systems**
- **Repair the microphone or power intermittency in the pickup truck system.**

CAPABILITY 6: MT63 SKILLS

Capability Summary:

MT63 is a fast digital keyboard-based and potentially file-based mechanism to send accurate broadcast (1-to-many) information that can be very effective in sending bullets and broadcast messages.

This capability was not utilized during this Exercise.

CAPABILITY 7: PACKET CHAT

Capability Summary: Packet Chat skills were hoped to provide a way for participants to allow multi-party typed (digital) discussion similar to what can happen on a voice radio frequency. These skills were tested by a small number of participants in Thursday evening packet roundtables associated with other ARES training nets, and were easily acquired by participants. However, the function itself on the digital repeaters was found to be easily overloaded, so the utility of this skill without higher speed "mesh" communications networks is questionable.

This capability was not utilized during this Exercise.

CAPABILITY 8: LINBPQ CHAT FUNCTIONS

Capability Summary: LINBPQ, the software employed in much of the digital infrastructure created in the amateur community locally in the last year, allows for a "roundtable" chat discussion, forwarding each person's typed comments to the others involved. Unfortunately, the limitations of 1200 Baud Packet AX.25 are that this is unwieldy for more than about 3 active participants. Although in the planning stages it was hoped to be a useful function, by the time the Exercise had arrived, it was already known that the technology has significant limitations and its use was not as strongly advocated, with alternatives over voice suggested.

This capability was not utilized during this Exercise.

CAPABILITY 9: ICS FORMS

Capability Summary: ARES volunteers have been becoming more accustomed to standard FEMA/ICS forms through efforts of Jeff Capehart at previous simulation events. It is desirable that they be familiar with personnel log in forms, and essential that they be familiar with communications logs and message formats, particularly ICS-213 ("general message").

> **Activity 9.1** Utilize ICS-205 frequency chart, ICS personnel log in forms, ICS communications logs, and transfer ICS-213 record traffic.
>
>> **Observation 9.1.1:** Strength. For this Exercise, over a dozen ICS forms were filled out and utilized.
>>
>>> **Analysis** Our group is becoming quite accustomed to ICS forms.
>>
>> **Observation 9.1.2: Weakness**. We were not uniform in using the Activity Form; personnel at COMM UNIT #3 were so preoccupied in repairing broken software that they were not able to fill in the Activity Form.
>>
>>> **Analysis** Now that we have become more familiar with these forms, have Operations section provide clipboards/Activity Form and emphasize keeping track of difficulties for better analysis afterwards.

Recommendation:

- Now that we have become more familiar with these forms, have Operations section provide clipboards/Activity Form and emphasize keeping track of difficulties for better analysis afterwards.

ADDITIONAL ISSUES/RECOMMENDATIONS

- A remarkably uniform difficulty experienced by all groups who were located out of doors, was difficulty in viewing computer screens. One volunteer strongly recommended that we be equipped with "photographer's capes"' another wanted collapsable, foldable cardboard shields so that screens could be better seen. We will need to find either of both of those solutions for future service.

SECTION 7: AFTER ACTION REPORT CONCLUSION

Operation ARES STEINHATCHEE STORM was conducted on October 7, 2017, to test Alachua County ARES capabilities to provide backup emergency communications to a simulated, hurricane-devastated community.

This was undoubtedly one of the most ambitious Full Scale Exercise ever carried out by Alachua County ARES, at least within known history. A very wide array of communications skills were put to the test, including simplex VHF voice, simplex VHF repeater, duplex VHF repeater, HF WINLINK, VHF WINLINK, microwave WINLINK, microwave VoIP.

These digital skills (WINLINK, packet, HF) are still new to our local volunteers, and this exercise solidified their usage of them. We added the new mode of Microwave for this Exercise --- and the deployment 60 miles out of our home town. Furthermore, we strenuously tested their abilities to emplace emergency antennas and provide alternative power --- all things that would be important in a real hurricane / communications emergency. Our group performed admirably at all these tasks. The real performance issues were external to our group – the WINLINK server software experienced difficulties perhaps related to a Microsoft Windows update, and our hoped-for VHF link from lookout towers 60 miles apart was not successful. We will need to have backup copies of all software for any deployments, and in order to reach Steinhatchee by VHF digital repeater we will need an intermediate station, perhaps at the Trenton look out tower. Over all, we discovered the HF system is more reliable and (once software works) easier to use than the multiple sequential connections required for VHF packet long distance connections. End-volunteers found the VHF-to-HF system very easy to use, and the Microwave-to-HF system also was easy to use. These were significant advancements for our group. This was the very first time that our group ever deployed a portable WINLINK RMS server system.

This exercise allowed us to begin to put together our own internal Incident Command System, which helped broaden our leadership involvement. For this exercise, message creation was moved to new leaders and worked well.

A weakness of our deployment was that we were unable to secure involvement by the Alachua County Emergency Operations Center; officials cited responsibilities still remaining from Hurricane Irma of just a few weeks previous. We were also unable to secure involvement by a nearby amateur club we have provided some training for. However, we were delighted to have volunteers from other groups including the Ocala Hospital Emergency Communications group, and the Black Swan organization attempt to assist us. Furthermore, we had cooperation from a newly-added WINLINK sysop, Ray Cook.

At our previous Full Scale Exercise, Larry Rovak commented that out top priority should be to keep "growing our infrastructure" --- and this Exercise did exactly that.

After Action Report / Improvement Plan (AAR/IP)　　　NF4RC　　　2017 Hurricane Test

APPENDIX A: ISSUES NOTED / IMPROVEMENT PLAN

(updates ongoing at: http://qsl.net/nf4rc)

Updated as of Oct 25 2017

No.	Issue	Suggestion	Actual Action Taken
1	**WINLINK RMS_RELAY could not connect to RMS_TRIMODE, fatally crippling the BRIDGE portal to automated HF forwarding.** Not clear exactly why; a Microsoft Windows update may have been involved; a flaw in TRIMODE requiring internet access that was repaired contemporaneously by the Winlink Development Team may have been involved; the ARES volunteer did not bring backup installation software, so a new installation had to be downloaded over the Internet (violation of exercise plan) but the system then worked well.	1. NEVER DEPLOY WITHOUT FLASH DRIVES WITH RE-INSSTALLATION SOFTWARE. 2. Investigate moving server software to linux-based bpq systems. 3. When using microsoft-windows based softwre, sequester equipment from last testing to deployment time and prevent either Microsoft or Winlink upgrades.	**Immediate problem RESOLVED** 1) re-installation during the exercise cured the immediate problem. 2) Winlink upgrades are said to have removed the TRIMODE internet-absent startup-issue. (This needs to be tested.) **Longer Term Solutions:** 1) Efforts underway to investigate moving to linux-based server platform. 2) As an immediate improvement, a Windows 8.1 laptop has been obtained (no WINDOWS updates after January 2018) and software installed; successfully runs system.
2	VHF link from Jonesboro Lookout Tower to NEWB did not reliably connect. Furthermore, station LKCTY was not reachable on SEDAN frequency.	1. We would need an intermediate link, possibly at Trenton tower, to make this work. 2. Unclear the status of LKCTY which had a very	UNRESOLVED – but I think we can get permission for the Trenton tower, and that might make a significant improvement for VHF coverage

After Action Report / Improvement Plan (AAR/IP)   Alachua County ARES   2017 Hurricane Test

		strong signal during pre-exercise testing.	
3	The only station with excellent VHF voice connection to the far TOWER comm unit #4, was the 40-watt GE transceiver utilized by Comm Unit #1.	Work toward putting 40-watt units in all deployed comm units. And or find some little 3 element beams?	**RESOLVED:** EOC/ARES have created 3 higher power go-boxes – that helps resolve this. I have created one higher power go-box myself also.
4	Outdoor units were unable to easily see their screens.	Find some form of shading – photographer's capes made from black fabric, or folding / collapsible cardboard or metal shields.	**UNRESOLVED—we need to cut some cardboard and get some velcro and build some of the adjustable screens.**
5	NEED MORE INK PENS		**RESOLVED:** I'll ask Operations Chief of future events to handle this.
6	Possibly a better sign up sheet that more clearly shows where people are being sent.		The stock ICS-211 might be better adapted for our purpose. I'll work on that.
7	Need a session where people learn more about each others' radios so we can operate them better when needed.		**RESOLVED:** schedule for the JANUARY ARES meeting.

After Action Report / Improvement Plan (AAR/IP)　　NF4RC　　2017 Hurricane Test

APPENDIX B: SELECTED ICS FORMS UTILIZED

ICS 201 SITUATION

ICS 201 INCIDENT BRIEFING　　1030 LOCAL TIME
INCIDENT: EXERCISE STEINHATCHEE STORM
(5)　　SITUATION SUMMARY

1030 (Local Time) SITUATION

<u>Incident Command Post has been established at the Who Dat Bar & Grill, at the Good Times Marina and Motel on the South side of Steinhatchee River, roughly opposite Sea Hag Marina.</u>

CHANGES SINCE 0900 BRIEFING:
Cruise Ship PoorNavigator moved into Deadman Bay by winds and steering malfunctions, has grounded solidly in 9-10 feet of water approximately 4 miles NW of Marker #1. Ship's draft is approximately 19 feet. Listing 30 degrees to port. Motive power lost; most on-board power lost; water supplies unpressurized; multiple injuries not yet documented; 600 total crew+passengers on board. Communications via Marine FM radio, channels 21,68,69.

Winds now 30 mph; waves still 4-5 feet in Deadman Bay; expected to decline to 3feet by Noon, winds decreasing to 20 knots at Noon; Rain Level 2 at present, expected to slacken.

Marking of submerged hazards in Steinhatchee river to the mouth has begun.

Approximately 20 flat bottom craft have been salvaged between Sea Hag, Good Times, and River Haven marinas. Refueling operations in progress in preparation for rescue efforts. Cruise ship is working on egress points, ladders, slides. Waves/Winds are still prohibitive for rescue.

Communications
Volunteer amateur radio team is expected to arrive shortly, is tasked with sending urgent new resource requests by way of Alachua County EOC/ State EOC. Status updates to Alachua County8 EOC/State EOC planned at 45 minute intervals.

Crews have opened roads to Perry hospital.

ASSIGNMENT LIST (ICS 204)

1. Incident Name: EXERCISE STEINHATCHEE STORM

2. Operational Period:
- Date From: OCT 7
- Time From: 1030
- Date To: OCT 7
- Time To: 2100

3.
- Branch: LOGISTICS SERVICE
- Division:
- Group:
- Staging Area: CADES COVE

4. Operations Personnel: Name — Contact Number(s)
- Operations Section Chief: (LOGISTICS) JOHN TROUPE
- Branch Director:
- Division/Group Supervisor:

5. Resources Assigned:

Resource Identifier	Leader	# of Persons	Contact (e.g., phone, pager, radio frequency, etc.)	Reporting Location, Special Equipment and Supplies, Remarks, Notes, Information
BRIDGE	GORDON GIBBY	2	146.52 fm SIMP / 145.07 AX.25	BRIDGE; HF/MICRO/VHF NODECO
HUNGRY HOWIES	SUSAN HALBERT	2	146.52 FM SIMP / MICROWAVE	HUNG HOW; VHF/MICROWAVE
GOOD TIMES	VANN CHESNEY	2	146.52 FM SIMP / 145.07 AX.25	ROYS REST; VHF DIGITAL
LOOKOUT	ART GRANT	2	146.52 FM SIMP / 145.07 AX.25	JONESBORO LOOKOUT TOWER

6. Work Assignments:

[ ] Comm. Sup. Team #1 (NODE) - Establish VHF Node NF4RC-2 / HF Forwarding NF4RC / Microwave Access, at Bridge; provide relay services, messages as requested.

[ ] Comm. Sup. Team #2 (Micro) - Establish Microwave access at Hungry Howies KG4VWI. OBTAIN LIST OF PERSONS ON CRUISE SHIP POOR NAVIGATOR FROM EOC/CRUISE LINE.

[ ] Comm. Sup. Team #3 (VHF) - Establish VHF digital AC4QS @ WHO DAT BAR & GRILL, GOOD TIMES MARINA. Commencement message to Alachua EOC; other messages as requested. SEND RESOURCE REQUESTS FROM INCIDENT COMMANDER & STATUS UPDATES TO EOC / STATE EOC.

[ ] Comm. Sup. Team #4 (NODE) - Establish VHF Node NF4RC-4 at Jonesboro Tower, provide relay services.

7. Special Instructions:

MAINTAIN ROAD SAFETY AT ALL TIMES. USE SUN-COVER AND SUNSCREEN. STAY HYDRATED.

8. Communications (radio and/or phone contact numbers needed for this assignment):

Name/Function	Primary Contact: Indicate cell, pager, or radio (frequency/system/channel)
JEFF CAPEHART - INC. COMM.	146.52 FM SIMPLEX; 145.070 AX.25
JOHN TROUPE - LOG. CHIEF	146.52 FM SIMPLEX; 145.070 AX.25

9. Prepared by: Name: GORDON GIBBY Position/Title: PLANNER Signature:

ICS 204 | IAP Page: 3 | Date/Time: 8/24/2017 1900

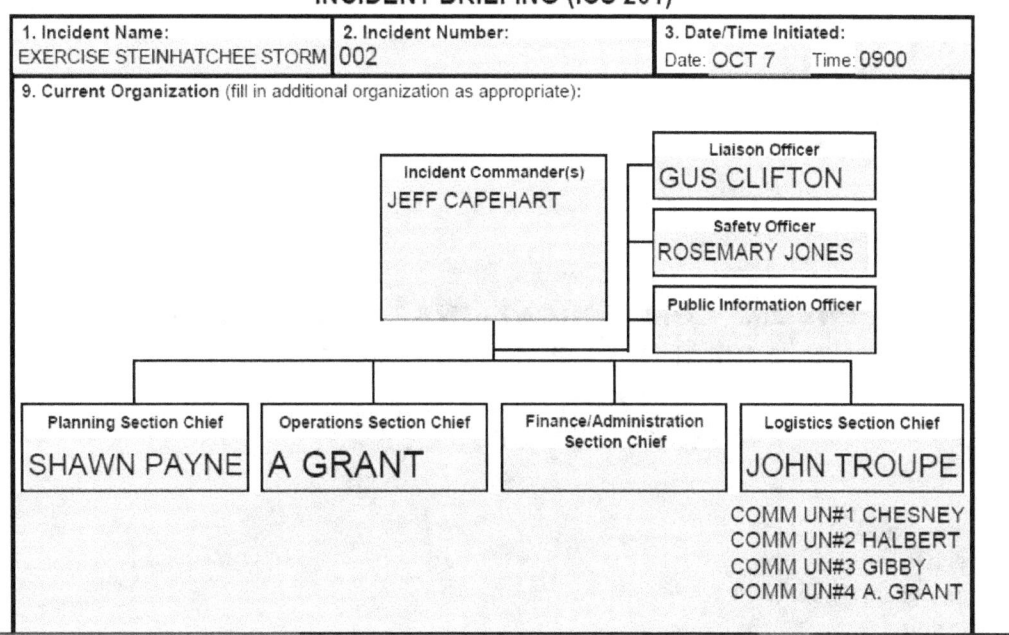

APPENDIX C: ICS 205A ADDENDUM

ADDENDUM: COMMUNICATIONS LIST (TO ICS 205A)
EXERCISE STEINHATCHEE STORM
FINAL REVISION
THURSDAY OCTOBER 5 2017

NOTE:
Dixie County Sheriff Dept
(notified and aware of our Exercise)
Captain John Simmons
352 578 4128

1. Incident Name: **EXERCISE STEINHATCHEE STORM**	2. Operational Period	Date From Oct 7 Date to: Oct 7 Tie from: 0900 Time to: 2100	
3. VHF / HF / Microwave Communications Layout:			
Incident or other Position	Call / Alias / Function	Frequency	Details
			NOTE that the comm units are laid out in order from west to east 1 = Good Times (IC) 2 = Hungry Howies 3 = Bridge 4 = Lookout Tower, and that as much as possible, SSID's and other numerical identifies are harmonized.

After Action Report / Improvement Plan (AAR/IP) NF4RC 2017 Hurricane Test

GOOD TIMES Incident Command Post (ICP) COMM UNIT 1, Vann Chesney Leader.	BAND	CALL	Voice: 146.550 AX.25: 145.070	BAND	EQUIPMENT
	HF			HF	
	VHF	AC4QS **GOOD TIMES**		VHF	Vann's go-box (battery) + voice walkie talkies
	Micro			MICRO	

HUNGRY HOWIE Logistics Chief COMM UNIT 2 Susan Halbert Leader NOTE: please do not make any changes to any ip numbers as it can change the entire network address assignment.	BAND	CALL	Voice: 146.550 Micro: 2.397G 5 MHz BW (if any additional gear, AX.25 145.070)	BAND	EQUIPMENT
	HF			HF	
	VHF	KG4VWI **HUNGRY HOWIES**		VHF	voice walkie talkies
	Micro	NF4RC-002 VOIP2		MICRO	Cubical microwave go-box with Nanostation (battery) KX4Z-002 (root: ONETIME) WIFI IP 10.254.254.2

POP3 username/ passwords for RMS_RELAY pop3 access

KG4VWI@WINLINK.ORG: (ONETIME)

NK3F@WINLINK.ORG: (ONETIME)

Additional username / passwords can be created as needed.

ETHERNET LAN
DHCP by Ubiquiti

Ethernet Device	IP
Nanostation	10.247.240.17
VOIP2 00:0b:82:af:26:c5	10.247.240.19
Available	10.247.240.18 10.247.240.20
LENOVO 50:7b:9d:02:db:e7	10.247.240.22

BRIDGE AREA COMM UNIT 3 Gordon Gibby Leader NOTE: please do not	BAND	CALL	HF: Multiple frequencies, both automatic and manual. 2 meters: Voice: 146.550 AX.25 145.070 Micro: 2.397 Ghz, 5 MHz BW	BAND	EQUIPMENT
	HF	NF4RC		HF	ICOM HF large cube go-box
	VHF	NF4RC-3 *BRDGE* **BRIDGE**		VHF	ICOM 229 cube go-box
	Micro	NF4RC-003 VOIP3		Micro	Flat Board Microwave with Ubiquiti Bullet NF4RC-003 (root: ONETIME) VOIP 3

53

After Action Report / Improvement Plan (AAR/IP)   Alachua County ARES   2017 Hurricane Test

make any changes to any ip numbers as it can change the entire network address assignment.				WIFI: 10.254.254.3		
				Ethernet LAN		
				Ethernet Device		IP
				Bullet (LAN) (DHCP)		10.247.240.25 (http: on 8080)
				VOIP3 00:0b;82:af:26:d0		10.247.240.26
				NF4RC-3 b8:27:eb;11:2f:6a		10.247.240.27
				NF4RC RMS LENOVO computer **WINLINK telnet access: 10.247.240.29 Port 8772** **Email Access:** SMTP: 25 POP3: 110		10.247.240.29
				WEB SERVER RASPBERRY		10.247.240.30: 8080 (not yet reserved
				SPARE		10.247.240.29
				Units powered from Truck Battery		
Jonesboro Lookout Tower **COMM UNIT 4** **Art Grant, Leader**	BAND	CALL	Voice: 146.550 AX.25: 145.070	Band		Equipment
	HF			HF		
	VHF	NF4RC-4 ***LKTWR*** LOOKOUT TOWER		VHF		Art go box + Icom 28 go-box + (gell cell) + walkie talkies
	Micro			Micro		
DISTANT CONTACTS						

Forest Grove	NF4RC-7 / NEWB digital AX.25 node	AX.25 145.070	RELAY NODE
WINLINK GATEWAY KX4Z-7	KX4Z-7 / GNVWLK digital AX.25 node/ WINLINK Vhf/HF forwarding	AX.25 145.070 AX.25 145.030	WINLINK TERMINAL
WINLINK GATEWAY NK3F-7	NK3F-7/10 digital AX.2 node/ internet gateway	AX.25 145.070 AX.25 145.770	WINLINK TERMINAL
WINLINK GATEWAY KM4YGH-7/10	KM4YGH-7/10 digital AX.25 node / internet gateeway	AX.25 145.070	WINLINK TERMINAL
ALTERNATE ROUTE (requires switching everyone to 145.770)			
Lake City SEDAN	KE4BQI-? LKCTY 145.770		Can then transition to GARC2 (145.770), to NK3F-10 (145.770), Other options may also exist
TERMINAL CONTACTS	ACTUAL CONTACTS (DO NOT USE UNLESS EMERGENT)	**PROXY For 2017 Simulated Emergency Test**	COMMENT

After Action Report / Improvement Plan (AAR/IP)   Alachua County ARES   2017 Hurricane Test

Alachua Cty. EOC Jeff Bielling	Main Number XXX-XXX-XXXX JEFF BIELLING PERSONAL PHONE (emergencies only) XXX-XXX-XXXX To send text, email to: (VERIZON) XXXXXXXXX@vtext.com	**Email:** send to both <u>docvacuumtubes@gmail.com</u> and simultaneously to <u>KX4Z@WINLINK.ORG</u> **Packet keyboard contact:** Nancy KM4YGI will be on 145.070 and reachable via KX4Z-7 on 145.070 To buzz her phone: (ATT) send email to XXXXXXXX@txt.att.net **Packet YAPP FILE TRANSFER:** connect to KM4YGI (auto-receive) on 145.070	For the Exercise, be sure to put TEST EXERCISE first thing in message nd be sure your subject auto starts with //WL2K
COAST GUARD "APPALACHICOLA" FICTITIOUS SERVICE		(Ohio ham group) XXXXXXXXXX COM-L AuxComm Emergency Coordinator, Hamilton County, Ohio ARES EMAIL: XXXXXXXXXX Cell phone: XXXXXXXXXXX	For the Exercise, be sure to put TEST EXERCISE first thing in message nd be sure your subject auto starts with //WL2K Bryan is going to be simultaneously helping provide communications for a Marathon, so don't overburden him with too much to do!
COAST GUARD "TAMPA" FICTITIOUS SERVICE		(Sarasota ham group) EMAIL: XXXXXXXXXXXX	For the Exercise, be sure to put TEST EXERCISE first thing in message and be sure your subject auto starts with //WL2K

		Cellphone: XXXXXXXXXXX (Verizon). To text his cellphone, send winlink email to XXXXXXXX@vtext.com (You'll be reaching XXXXXXX Sarasota Digital Grouop Sarasota County Auxiliary Communications Service XXXXX@comcast.net)	
State Watch Officer	swp@em.myflorida.com emergency: **800-320-0519** Non Emergency: **850-815-4001**	XXXXXXXXXXX XXXXXXX@WINLINK.ORG and he wants THIS IS A DRILL first in the body And he wants to be addressed as: XXXXXXX, Acting Hospital Incident Commander, I Medical Center, Ocala Fl.	*STATE OF FLORIDA EMERGENCY MANAGEMENT email observed 24 hours per day*
State EOC		XXXXXXXXXX EMAIL: XXXXXXXX@gmail.com Phone: XXXXXXXXXX	For the Exercise, be sure to put TEST EXERCISE first thing in message and **be sure your subject auto starts with //WL2K**
NOAA Rescue Coordination Centers	Main phone line: (907) 551-7230 Toll free: 1-800-420-7230		*Consider using eFAX to send a fax by email to their FAX number. You must have created an account beforehand... there is a charge.*

	FAX; (907) 551-7245 MIAMI: D07-SMB-CMDCENTER@USCG.MIL NORFOLK: D05-SMB-D5CC@uscg.mil USA: afrcc.console@us.af.mil		https://www.efax.com/how-it-works
Gilchrist County EOC	386 935 5400		
Dixie County EOC	352 498 1240		17600 SE Highway 19, Cross City, FL 32628
Taylor County EOC	850 838 3575		591 E. US Highway 27, Perry, FL 32347 stephen.spradley@taylorcountygov.com (Steve Spradley, EM Director)
Shands Hospital Anesthesia Call (north)	XXXXXXXXXX text: sprint		carried 24 hours per day
Shand Hospital Anesthesia Call (south)	XXXXXXXXXX text: sprint		carried 24 hours per day
Gainesville Airport University Air Center FBO	352 335 4681		
ShandsCair Emergency Dispatch	XXXXXXXXX		

Technique to use EMAIL to create a TEXT MESSAGE to a Cell Phone, if you know the CARRIER COMPANY of that Cell Phone:	• AT&T: number@txt.att.net • T-Mobile: number@tmomail.net • Verizon: number@vtext.com (text-only), number@vzwpix (text + photo) • Sprint: number@messaging.sprintpcs.com

	or number@pm.sprint.com • Virgin Mobile: number@vmobl.com • Tracfone: number@mmst5.tracfone.com • Metro PCS: number@mymetropcs.com • Boost Mobile: number@myboostmobile.com • Cricket: number@mms.cricketwireless.net • Ptel: number@ptel.com • Republic Wireless: number@text.republicwireless.com • Google Fi (Project Fi): number@msg.fi.google.com • Suncom: number@tms.suncom.com • Ting: number@message.ting.com • U.S. Cellular: number@email.uscc.net • Consumer Cellular:number@cingularme.com • C-Spire: number@cspire1.com • Page Plus:number@vtext.com

APPENDIX D: LESSONS LEARNED

While the After Action Report/Improvement Plan includes recommendations which support development of specific post-exercise corrective actions, exercises may also reveal lessons learned which can be shared with the broader homeland security audience. Federal Emergency Management Agency (FEMA) maintains the *Lessons Learned Information Sharing* (LLIS.gov) system as a means of sharing post-exercise lessons learned with the emergency response community. This appendix provides jurisdictions and organizations with an opportunity to nominate lessons learned from exercises for sharing on *LLIS.gov*.

For reference, the following are the categories and definitions used in LLIS.gov:

- **Lesson Learned:** Knowledge and experience, positive or negative, derived from actual incidents, such as the 9/11 attacks and Hurricane Katrina, as well as those derived from observations and historical study of operations, training, and exercises.

- **Best Practices:** Exemplary, peer-validated techniques, procedures, good ideas, or solutions that work and are solidly grounded in actual operations, training, and exercise experience.

- **Good Stories:** Exemplary, but non-peer-validated, initiatives (implemented by various jurisdictions) that have shown success in their specific environments and that may provide useful information to other communities and organizations.

- **Practice Note:** A brief description of innovative practices, procedures, methods, programs, or tactics that an organization uses to adapt to changing conditions or to overcome an obstacle or challenge.

Exercise Lessons Learned

The following subject headings are lessons derived from the Alachua County, Florida FSE on May 6, 2017 that are proposed for inclusion in the Department of Homeland Security's Lessons Learned/Best Practices web portal, LLIS.gov:

- VHF and microwave communications worked well.

- THE ASSISTANCE OF STATE, LOCAL, AND PRIVATE ENTITIES CONTRIBUTED GREATLY TO THE LEARNING OPPORTUNITIES AFFORDED BY OUR EXERCISE.

APPENDIX E: PARTICIPANT FEEDBACK SUMMARY

PARTICIPANT FEEDBACK FORM

(SUGGESTED FOR USE IN SUBSEQUENT EXERCISES)

Exercise Name: _____   Exercise Date: _____

Participant Name: _____   Title: _____

Agency: _____

Role: __Player __Observer __Facilitator __Evaluator

PART I: RECOMMENDATIONS AND CORRECTIVE ACTIONS

1. Based on the exercise today and the tasks identified, list the top 3 strengths and/or areas that need improvement.

   _____

   _____

   _____

2. Is there anything you saw in the exercise that the evaluator(s) might not have been able to experience, observe, and record?

   _____

After Action Report / Improvement Plan (AAR/IP)   Alachua County ARES   2017 Hurricane Test

_____
_____
_____

3. Identify the corrective actions that should be taken to address the issues identified above. For each corrective action, indicate if it is a high, medium, or low priority.

_____
_____
_____
_____

4. Describe the corrective actions that relate to your area of responsibility. Who should be assigned responsibility for each corrective action?

_____
_____
_____
_____

5. List the applicable equipment, training, policies, plans, and procedures that should be reviewed, revised, or developed. Indicate the priority level for each.

_____
_____
_____

PART II – EXERCISE DESIGN AND CONDUCT: ASSESSMENT

Please rate, on a scale of 1 to 5, your overall assessment of the exercise relative to the statements provided below, with **1** indicating **strong disagreement** with the statement and **5** indicating **strong agreement.**

Table C.1: *Participant Assessment*

Assessment Factor	Strongly Disagree				Strongly Agree

a.	The exercise was well structured and organized.	1	2	3	4	5
b.	The exercise scenario was plausible and realistic.	1	2	3	4	5
c.	The facilitator/controller(s) was knowledgeable about the area of play and kept the exercise on target.	1	2	3	4	5
d.	The exercise documentation provided to assist in preparing for and participating in the exercise was useful.	1	2	3	4	5
e.	Participation in the exercise was appropriate for someone in my position.	1	2	3	4	5
f.	The participants included the right people in terms of level and mix of disciplines.	1	2	3	4	5
g.	This exercise allowed my agency/jurisdiction to practice and improve priority capabilities.	1	2	3	4	5
h.	After this exercise, I believe my agency/jurisdiction is better prepared to deal successfully with the scenario that was exercised.	1	2	3	4	5

PART III – PARTICIPANT FEEDBACK

Please provide any recommendations on how this exercise or future exercises could be improved or enhanced.

_____

_____

_____

_____

_____

_____

_____

APPENDIX F: EXERCISE EVENTS SUMMARY TABLE

Table D.1: *Exercise Events Summary*

Date	Time	Scenario Event, Simulated Player Inject, Player Action	Event/Action
10/07/17	0700-0730	Participants gathered at County Foodly for Breakfast	Discussion.
10/07/17	800	Deploy caravan style to Stienhatchee using 146.55 Simplex.	Travel
10/07/17	1000	Check-In at Casey's Cove	ICS-211 checkin
10/107/17	1030	Deploy to locations, establish communications, begin transferring messages	Message were pre-developed by the Incident Commander and Logistics Chief.
10/07/17	1200	Move toward Lunch and demobilization procedures	Discussion and feedback

APPENDIX G : ACRONYMS

Acronym	Meaning
AAR	After Action Report
ALS	Advanced life support
CDC	Centers for Disease Control and Prevention
DHS	Department of Homeland Security
EDS	Emergency Dispensing Site
EMA	Emergency Management Agency
EMS	Emergency Medical Services
FEMA	Federal Emergency Management Agency
FOUO	For Official Use Only
FPC	Final Planning Conference
HF	High Frequency (shortwave)
HSEEP	Homeland Security Exercise and Evaluation Program
IAP	Incident Action Plan
IC	Incident Commander
ICS	Incident Command System
IC/UC	Incident Command/Unified Command
IPC	Initial Planning Conference
LLIS	Lessons Learned Information Sharing
MDPH	Massachusetts Department of Public Health
MEMA	Massachusetts Emergency Management Agency
MPC	Midterm Planning Conference
MRC	Medical Reserve Corps
MSEL	Master Scenario Events List
NIMS	National Incident Management System
POC	Point of contact
RSS	Receipt, Stage and Storage facility
SARNET	Statewide Amateur Radio Networking (a connected series of amateur radio repeaters)
SNS	Strategic National Stockpile
TCL	Target Capabilities List
UC	Unified Command
VHF	Very High Frequency (30-300 MHz)
WINLINK	A radio email system, see www.winlink.org

APPENDIX H: RMS_RELAY SETTING

Disclaimer: I cannot claim to be an expert on RMS_RELAY settings, which are often complex and their documentation more difficult to find. The opinions given here are merely my best suggestions for newer users of RMS_RELAY. Mike Burton, and others of the Winlink Development Team have helped me many times....and even they seemed confused at times.

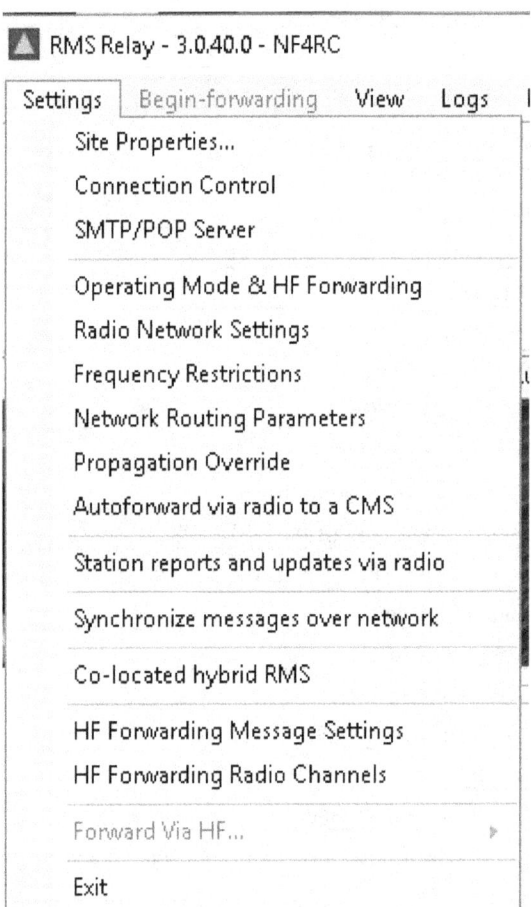

The most troublesome settings are the:
- Operating Mode & HF Forwarding
- Radio Network Settings
- Frequency Restrictions
- Autoforward via radio to a CMS

Operating Mode and HF Forwarding Control

Operating Mode
- ○ Only accept connections if Internet is available
- ○ If Internet is not available, accept only radio-only forwarding connections and deny CMS connections
- ○ Radio-only, local message hub -- Store messages locally. Do not upload messages through Internet
- ● Hold for Internet access -- Store messages locally until Internet is available, then upload them
- ○ Stand-alone Network "post office" - Not part of the Winlink network

HF Message Forwarding Control
- ○ Do not forward messages via HF
- ○ Forward messages via HF to another RMS connected to the Internet
- ● Operate as a node in the Winlink hybrid network (Trimode must run)

Minutes to delay after Internet loss before starting radio-only network operation: `2`

Automatic Sending Control
- ☑ Enable automatic operation
  - Seconds before starting: `20`
  - Minimum seconds between sends: `60`
  - Maximum minutes sending: `20`
- ☑ Check for busy channel before transmitting
- ☐ Emphasize Pactor signals for busy detection
  (Requires P4 modem with 1.17.8 or later firmware)
  - Ignore busy after this many minutes: `600`
- ☑ Simulate Internet Outage

Trimode Control
- ☑ Automatically start and stop Trimode
- ☐ Start Trimode minimized

Folder where Trimode is stored:
`C:\RMS\RMS Trimode\`

IP: `127.0.0.1`   Port: `8510`
(Default port is 8510)

Pactor Level for Forwarding
Minimum: `2`   Maximum: `3`

Folder for VOACAP (itshfbc)
`C:\itshfbc\`

[Save]   [Cancel]

The Settings above are to **force the station to do radio RF forwarding of emails**. There are at least two ways to accomplish this. This method puts the system in HYBRID mode, but then instructs it to "Simulate Internet Outage" (check box, bottom left corner). An alternative is to disable your WIFI or any other internet access.

CAUTION: You should definitely test your system with the Internet totally disconnected, and all the way from a turned-off computer, through full operation.   It is best to learn of problems BEFORE your local area has lost all network access....

Radio Network Settings

Parameters specified on this screen control the operation of RMS Relay when it is functioning as a station in the Hybrid Winlink Network.

Propagation and Routing Control Files
- Propagation matrix file date: 2017-10-05-09:49
- MPS file date: 2017-09-16-17:42
- Latest SFI: 87 @ 2017/10/05 12:53 (UTC)

[Make Propagation Matrix and Message Pickup Station Files]

☐ Automatically generate files every day
Time of day (hh:mm, 24 hour, local time): 02:25

[List users selecting this RMS as their MPS]
[List all user MPS selections]

Blocked RMS
Callsigns of RMS that should not be called
Specify one callsign on each line

Station Forwarding Options
Extra time to foward through RMS (minutes): 10
Radio-only end point. No forwarding to other RMS ☐
Disallow forwarding to a CMS ☐

Notification when forwarding pending
Forward-pending sound: Asterisk   ☐ Repeat
E-mail/Text notification:

System diagnostic control
☐ Allow diagnostic information to be sent to the Winlink Development Team

[Save] [Cancel]

Radio Network Settings: The key point here is that "Automatically generate files every day" is UNCHECKED for this temporary usage exercise station --- because otherwise, when you first turn it on, it is going to start creating the Propagation Matrix!!! That will make the adjustment of RMS_RELAY almost impossible for 20-60 minutes depending on the speed of your computer.

Station Forwarding Options: I recommend allowing 10 minutes "Extra time to forward through RMS" --- but I'm not certain this makes any difference.

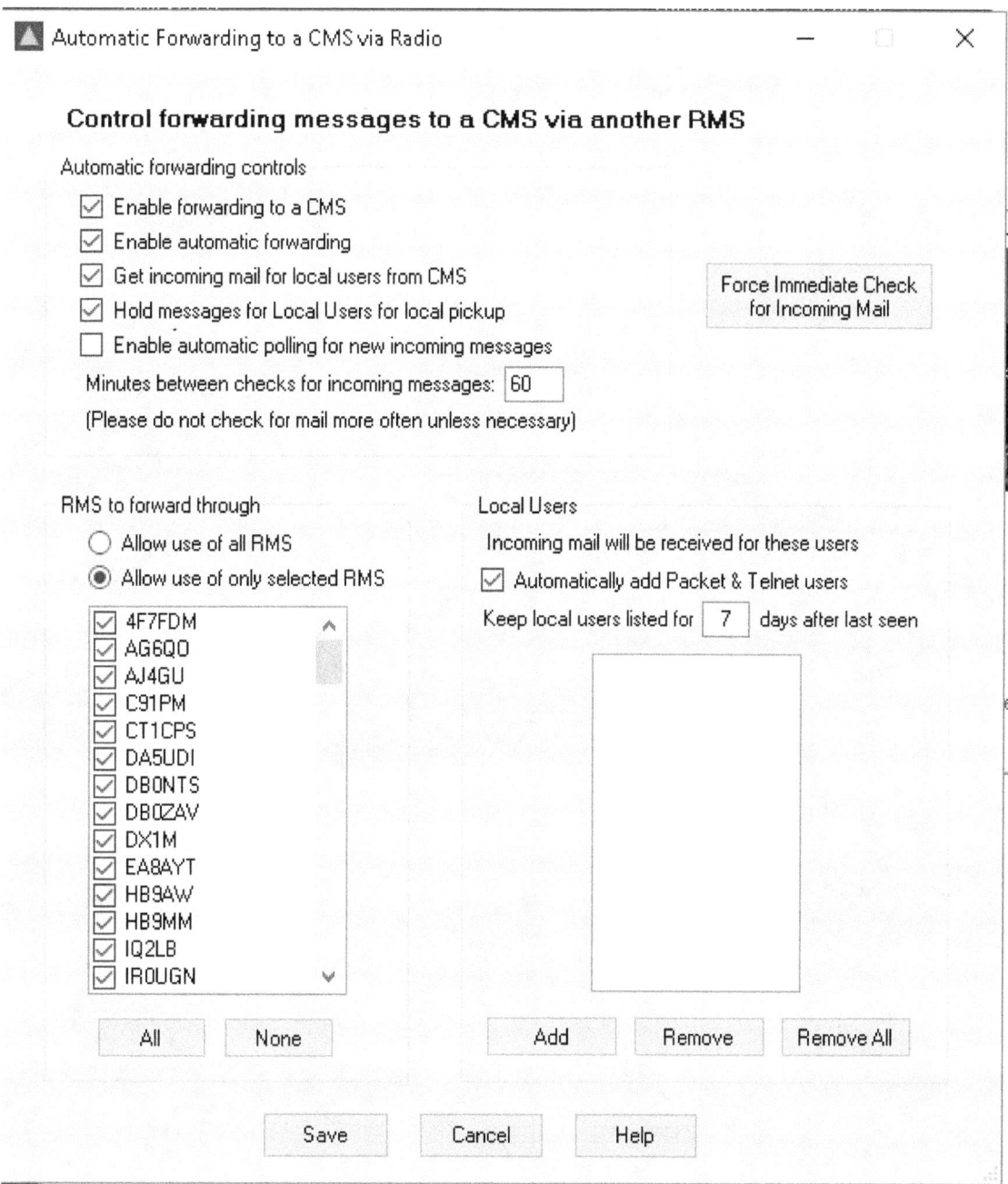

Automatic Forwarding Controls: The CMS Connections page is a tricky one and I do not completely understand the functionality; WINLINK forums often have murky problems here. Suggest that you check at least the 4 boxes checked in the upper left. If you wish POLLING (periodic automatic HF calls to check for any incoming mail) then check that box also and set maybe ever 45-90 minutes. Checks for incoming mail are automatically performed any time you have OUTGOING mail connections.

Although this illustration shows "Allow use of only selected RMS" checked, you should use this only if you have tested and found the best RMS's to AT THE TIME OF DAY OF YOUR NEED --- because it changes hourly with propagation. Otherwise, it may be wiser to choose "Allow use of all RMS."

Often, this list of RMS possibilities will not fill in. You may need to go in or out of "hybrid mode" – or some other action that causes a re-creation of the propagation matrix. **Be warned, that can take up to an hour, and may make it impossible to make other changes until this cpu-intensive task is completed.**

www.ingramcontent.com/pod-product-compliance
Lightning Source LLC
Chambersburg PA
CBHW082358220526
45470CB00008B/2787